The Human Spirit

How Science, Art, and Religion Shape Happiness

Hamilcar

Global East-West (London)

Global East-West (London).

Contents

Publisher's Preface:
The Human Spirit of Hamilcar

It is with immense honour and deep admiration that we at Global East-West Ltd present The Human Spirit: How Science, Art and Religion Shape Happiness, a masterful work by Hamilcar. The author's sharp pen and penetrating mind continue to shed light on the intricacies of the human soul. Completed in Paris in 2000, this book only found its way to publication in 2025 after a long journey reflecting the patience and perseverance of its creator. As publishers, we are honoured to unveil this timeless reflection, which serves as a mirror to our contemporary era, where the quest for happiness remains as universal as it is elusive.

Hamilcar, a pseudonym chosen for personal or professional reasons, is a writer and thinker who has already made his mark with works such as Art and Society: A Sociological Perspective on Creativity and the poignant novel The Silent Scourge, both of which were published under our imprint. Living in France, where he teaches English at a secondary school, Hamilcar draws on his dual vocation as an educator and author to explore the paradoxes of existence with unique sensitivity. L'Esprit Humain is the product of this sensitivity: a profound meditation on how the three pillars of civilisation — science, art, and religion — influence our understanding of happiness and our quest for meaning.

From the very first pages, Hamilcar confronts us with the illusion of perfect happiness, a chimera that haunts modern society. In the chapter entitled 'The Illusion of Perfect Happiness', he dissects with relentless lucidity how unrealistic expectations fuelled by the media and social networks trap us in a whirlwind of insatiable desires. "We are thus drawn into a relentless whirlwind of insatiable desires, our thirst for external validation exacerbating our futile quest for absolute fulfilment," he writes, accurately capturing the resulting existential void. Though written a quarter of a century ago, this reflection resonates with disturbing acuity in today's world, where instant gratification and 'virtual masquerades' dominate our interactions.

However, The Human Spirit does not merely diagnose our malaise; it also offers a path to redemption. Hamilcar invites us to untie 'the knots of our illusions' and embrace the simplicity of existence, rediscovering the raw beauty of the present moment. This radical awareness requires courageous introspection and constant questioning of our motivations. Only by exploring the roots of our aspirations can we transcend the ephemeral and grasp the timeless reality of our being. This message, with its almost timeless wisdom, echoes the great philosophical traditions while remaining firmly rooted in troubled modernity.

One of the most fascinating aspects of this book is the way in which Hamilcar intertwines science, art, and religion as complementary forces in our quest for happiness. In the chapter 'Science Conquers Mystery', he argues that great scientific revolutions often arise from imaginative audacity, breaking down the barriers of knowledge and opening up new horizons of understanding. He argues that this alliance between imagination and reason is at the heart of the most revolutionary advances, offering curious minds an opportunity to explore the universe through a new lens. This vision, which celebrates wonder in the face of original mystery, is a vibrant tribute to human curiosity.

Similarly, Hamilcar pays particular attention to art, seeing it as a fundamental expression of the human spirit. In 'The Sirens' Song of Art', he traces the evolution of artistic forms from the cave paintings of early hunter-gatherers to modern masterpieces, including Greek tragedies and medieval art imbued with the sacred. He writes that each artistic movement has helped shape our perception of the world, capturing the mysteries of life and the deepest emotions of the soul. Through painting, music, literature, and dance, art becomes a mirror of transcendence and a means of connecting with something greater than oneself.

Finally, religion — discussed in particular in When Faith Triggers Divine Laughter — is presented as both a potential refuge and a terrain for questioning. Hamilcar asks, 'Is the House of God a refuge?' This reflection, tinged with subtle irony, invites readers to consider the role of faith in the pursuit of joy and meaning while acknowledging the diversity of spiritual experiences throughout history. This nuanced approach, which avoids dogmatism, is emblematic of Hamilcar's thinking: he is always searching for truth beyond established certainties.

By publishing The Human Spirit, we are introducing readers not only to an intellectually profound work, but also to an author whose vision, although rooted in the early 21st century, speaks directly to the challenges of our time. The decision to publish this text in 2025, twenty-five years after it was written, stems from a desire to allow these ideas to mature and to give them time to find their audience in a world ready to hear their message. At Global East-West LTD, we firmly believe that Hamilcar's ideas, expressed in poetic yet incisive prose, have the power to inspire a new generation of thinkers and dreamers.

This book is also a testament to Hamilcar's commitment to exploring humanity in all its complexity. Through his reflections on the 'great misunderstanding' that clouds our senses and his chronicles of 'insatiable desire', he reminds us that the true strength of the mind lies in its capacity to embrace paradoxes and find solace in the infinite abstract.

This perpetual struggle against the absurd, which he describes with poignant beauty, lies at the heart of what he terms 'the inextinguishable essence of humanity'. As publishers, we are proud to protect the integrity of this work in accordance with the copyright that guarantees its exclusivity.

We hope that readers will find The Human Spirit to be not only a source of reflection, but also an invitation to rediscover their own humanity through science that broadens our horizons, art that touches our souls and spirituality that questions our place in the universe.

We wish you a transformative and enriching read — a journey to the heart of the human spirit guided by Hamilcar's unique vision.

Global East-West LTD

2025

1

The Illusion of Perfect Happiness

Myth or Reality?

The Dazzling Mirage of Absolute Fulfillment

In the modern age, humanity is engaged in a frenetic pursuit of happiness, frequently seduced by transient illusions that distort our perception of genuine fulfilment. The distorted echo of collective aspirations blends with the mesmerising whispers of the media and social networks, creating a complex web of unrealistic expectations. The subtle yet relentless pressure of society pushes us to pursue an illusory ideal defined by non-existent standards and artificial models. Thus, we find ourselves caught up in a relentless whirlwind of insatiable desires, with our thirst for external validation exacerbating our futile quest for absolute fulfilment. This dazzling illusion creates a distortion between our deepest aspirations and the ever-changing reality around us, generating a sense of existential emptiness. However, by closely examining these mirages, we discover that true happiness is not found in material accumulation or superficial recognition, but in the serene contemplation of our authentic selves. Contemporary societies, obsessed with instant gratification and virtual masquerades, have erected a monument to the ephemeral, distracting us from the essence of our humanity. In this dazzling light, we must untangle the knots of our illusions, embrace the simplicity of existence and rediscover the raw, profound beauty of the present moment. Liberation from illusion requires radical awareness, courageous introspection, and constant questioning of our motivations. By confronting our insatiable desires and exploring the roots of our aspirations, we can shed light on the dark twists and turns of our unquenchable quest for happiness. This allows us to transcend the fleeting illusion and grasp the timeless reality of our being.

Chronicles of Insatiable Desire: A Rush Towards Emptiness

The unfulfilled quest for absolute fulfilment breeds existential anxiety, haunting lost souls. Like an elusive mirage, perfect happiness continually eludes those who seek it frantically. Humanity has embarked on a frantic race towards an illusory goal, believing that ultimate satisfaction lies in the fulfilment of material and worldly desires. However, this insatiable desire drags people towards a devastating void, like an endless abyss.

These chronicles reveal the sad irony of a society focused on rampant consumption and the perpetual pursuit of fleeting pleasure. Each conquest becomes a new quest and each possession becomes a source of frustration. Beneath the veneer of abundance and prosperity, the human soul gradually loses its substance, sinking into a quagmire of unfulfilled desires.

This frantic rush towards emptiness raises fundamental questions about the nature of existence itself. What is the meaning of this perpetual quest? Why does humanity run relentlessly towards an ever-receding horizon? Isn't it time to reflect on true fulfilment, which lies beyond the artifice and illusions of modernity?

Rather than condemning this deep aspiration for fulfilment, we must question the paths taken to achieve it. The distress lies not in desire itself, but in how it is understood. The chronicles of insatiable desire call for a radical rethink of our values and priorities, and for us to take a good, hard look at ourselves and rediscover the essence of humanity.

These chronicles reveal the poignant paradox of insatiable humanity, torn between the aspiration for the absolute and the oppressive embrace of emptiness. This lucid observation demands deep reflection and collective awareness. As long as this blind rush towards emptiness continues, lasting fulfilment will remain an unattainable dream for those who cannot see beyond their fleeting desires.

The Twilight of Radiant Promises: Fall or Rebirth?

Torn between the fading remnants of promised bliss and the bitter realisation of growing disappointment, modern man is caught in an unbearable dilemma. The radiant promises of absolute happiness crumble like sandcastles in the relentless waves of human existence. Is this the inevitable twilight of our noblest aspirations, or an invitation to unexpected rebirth?

We scan the uncertain horizon, desperately searching for tangible landmarks in a perpetually changing world.

Faced with the disintegration of ideals once held dear, we must re-evaluate our notions of happiness and satisfaction. The twilight of bright promises signifies not only a dizzying descent into the abyss of disenchantment, but also an opportunity to reinvent our relationship with joy and contentment.

At the heart of this existential storm, a striking portrait emerges of the human soul in perpetual search of meaning and fulfilment. The collapse of bright promises reveals the fragility of our ideal constructs, but also offers hope for profound, intimate rebirth. Like a phoenix rising from the ashes of illusion, this period of turmoil could herald an era of genuine fulfilment, grounded in reality rather than illusion.

Although the fall of seemingly immutable ideals may engulf all hope, it also dispels the opaque mists that cloud our vision. In this revealing twilight, a glimmer of authenticity emerges, transcending the superficiality of vain aspirations. Consciousness, sharpened by the lucidity of disillusionment, is imbued with new clarity, paving the way for a rebirth that transcends the mirage of perfect happiness.

In this twilight of radiant promises, where light flickers between shadow and clarity, a delicate thread of profound reflection on the nature of our quest for happiness is woven. Between fall and rebirth, disillusionment and renewed hope, humanity is at a critical crossroads, with the future to be shaped by each individual's choices. The time has come to decide whether twilight will be the inevitable prelude to decline or the catalyst for a triumphant rebirth, nourished by the wisdom gained in darkness.

Beyond the Broken Mirrors: Reflections on a Modern Utopia

In a society that is perpetually searching for illusory perfection, we are often captivated by the image of an ideal reality: a modern utopia shaped by our hopes and emotions. However, as we journey through this distorting mirror of our ambitions, we are confronted with the painful truth of our dreams' intrinsic fragility. Our ardent desire for a flawless existence collides with the irregularities of everyday life and the constant duality of humanity. Thus, the modernist utopia, with its seductive promises, becomes an unattainable mirage: a fleeting illusion that vanishes at the slightest contact with tangible reality. Nevertheless, it is important to recognise that within this shattered splendour lies a unique beauty. It is in the process of seeking and questioning these ideals that our humanity is forged. Precious glimpses of our authentic human nature emerge through the cracks in our modern utopia, shaped by resilience and the will to transcend our limitations. We realise that the quest for perfection requires acceptance of imperfection; that the pursuit of utopia can only flourish through recognition of the imperfect reality from which it stems. Thus, beyond the broken mirrors of our deepest aspirations, we find not disillusionment, but wisdom tinged with humility. It is in this encounter between the ideal and the real that the true essence of our existence takes shape, steeped in contradictions and wonderful paradoxes. Therefore, it is essential to contemplate this modern utopia not as a predestined failure, but as an indispensable catalyst for our spiritual evolution. Each of these fractures and shards of vanished desires should be seen as an invitation to deeper reflection and a more intimate exploration of our shared humanity. In this way, the broken mirror fulfils its transformative mission, inviting the soul to free itself from the chains of illusory perfection and embrace the infinite richness of imperfection.

Whispering Voices of Unspoken Melancholy

Tormented hearts languish in darkness, navigating the depths of the soul in search of fleeting light. Unspoken melancholy creeps into the cracks of everyday life, weaving its elusive web around wandering thoughts. Like a funeral ballet, it envelops spirits thirsty for gentleness, plunging them into an abyss of silent suffering.

Modern man, lost in the maze of consumerist society, suffocates under the weight of appearances and conventions. Every day, he dons his immaculate mask, hiding the torments that gnaw at his insides behind a fake smile. The lamentations of lost souls echo through cities numbed by the incessant tumult of modernity.

The result of a frantic quest for illusory perfection, diffuse guilt grips individuals, holding them prisoner to their own chimeras. In this macabre waltz, authentic aspirations hide behind the deceptive veil of external success, condemning the dreamers to a life of servitude.

Yet, at the heart of this inner turmoil, an unwavering whisper remains, like a gentle breeze caressing the dark night. Splinters of humanity escape the cracks in oppression, nourishing hope for an enlightened rebirth. Though fragile, these whispers reveal the true nature of being, reminding each lost soul of its original splendour before the madness of the world tarnished its purity.

In the face of this unacknowledged melancholy, there arises an urgent need for authenticity. The ultimate quest of every consciousness hungry for truth becomes finding one's way back to oneself, stripped of the trappings and artefacts of vanity. In this exhilarating quest, the voices of melancholy whisper as messengers of inner freedom, inviting humanity to break free from the chains of illusion and embrace the essential beauty of existence. It is time to let these voices resonate within us and give them the audience they need to liberate us in our search for peace and fulfilment.

Fragments of Anxiety: The Absence of the True Self in Synthetic Eden

In this synthetic Eden, where every smile is tinged with artifice and laughter rings hollow, the wandering soul finds itself lost in a maze of illusions. Modern man, hungry for external validation, creates a false identity by moulding himself to fit social norms, thereby stifling the authentic voice of his inner being. The virtual walls of this gilded cage reflect a manufactured happiness around him — a cruel illusion that promises fulfilment while cultivating an insatiable emptiness.

The frantic quest for fleeting pleasures creates an inner fracture: a complacency fuelled by artifice that masks essential loneliness. It is in this desert of pretence that the absence of the true self emerges, stifled under the excesses of a futile and superficial existence. This synthetic Eden then becomes a theatre of masks, where everyone adopts a puppet identity, terrified by the blank stares of others and forced to maintain a false coherence between their true self and the image they project. Amidst these fragments of anxiety, a crucial question arises: what is the real substance behind these simulacra?

This disturbing blindness seems to feed on a visceral fear of facing our authenticity and naked truth, unaltered by social filters. The desire to belong and conform to pre-fabricated norms leads to the insidious forgetting of what makes us unique. Thus, we are condemned to contemplate the deceptive splendour of a synthetic Eden, reduced to empty, insubstantial shadows ready to vanish at the first breath of authenticity.

The absence of the true self in the synthetic Eden is the bitter fruit of our refusal to explore the turbulent depths of our inner selves. We all wander like sleepwalkers, trapped in a mirage that refuses to ease our torment. Thus, within the confines of this false reality, the authentic quest for inner peace gradually fades, suffocated by the illusion of sterile, impersonal perfection. We tirelessly search for ourselves, reconnecting with our essence and seeking to transcend artifice in our ardent desire to reclaim our identity and free our true selves from the suffocating embrace of Synthetic Eden.

The Elusive Embrace of Eternal Return

It is in the relentless pursuit of happiness that we discover the timeless paradox of human existence. Like an elusive echo in the depths of consciousness, the concept of eternal return reveals itself as an inextricable part of the tormented mind. Time, that cruel impostor, weaves its elusive threads around our fleeting aspirations, imprisoning the soul in a nocturnal dance where promises of renewal mingle with perpetual disillusionment.

Deep within each of us lies the vertigo of eternal recurrence. The hours stretch out in a monotonous cycle, inexorably marking the passage of time towards an uncertain tomorrow. In this tangle of possibilities and irreversibilities, escape seems out of reach while the emptiness of nothingness fleetingly caresses our ephemeral dreams. Eternal return, like an elusive spectre, haunts our thoughts, shaking our deepest certainties and amplifying the eclipse of our identity.

Thus, the fires of introspection burn brightly when the soul is confronted with the subtle inexorability of the endless cycle. The inner journey becomes the tragic epic of those who aspire to transcend their condition, yet find themselves trapped in limbo, fuelled by uncertainty and the perpetual ebb and flow of time. Each step taken on this quest implicitly returns us to the source of perplexity, while the shadows of the past cast melancholy across the elusive horizon of tomorrow.

However, in this dark epic, humanity embraces the poignant beauty of despair. Torn from its synthetic Eden, the ephemeral being finds the vital impetus to rise above the pain of the absurd, and it is in this perpetual struggle that the inextinguishable essence of humanity is revealed. It is thus in the conflagration of paradoxes that the true strength of the wandering spirit is revealed, tirelessly seeking refuge in the abstract infinity of eternal return.

Reminiscences of the Great Common Misunderstanding

In the uncertain dawn of our era, we are caught up in a great misunderstanding that clouds our senses and undermines the very foundations of our humanity. Through the maze of our labyrinthine existences, we have woven a complex web of ambiguous perceptions, creating a veritable metaphysical chaos. The deafening clash between our deepest aspirations and tangible reality invariably places us at the heart of existential confusion.

This misunderstanding spreads insidiously, permeating every corner of our collective consciousness. Our intellectual wanderings seem to stumble upon deceptive mirages, creating an ocean of illusions that blurs the distinction between truth and falsehood. Over the years, this confusion has crystallised into pervasive alienation, creating deep dissonance within our society.

The insatiable quest for perfect happiness and the illusion of a flawless existence lull our minds with their enchanting song, distancing us from our true essence. This inner discord, often ignored or repressed, has become an unbearable burden, plunging humanity into a desperate search for illusory escape routes. Rather than elevating us, the insatiable thirst for absolute happiness invariably draws us into a whirlwind of perpetual dissatisfaction.

Yet at the heart of this great misunderstanding lies a silent call for clarity and redemption. By exploring these dark twists and turns, we may glimpse a ray of hope: a collective awareness that would transcend our fleeting illusions. Recognising this misunderstanding as part of the human experience could be the key to inner liberation, freeing our souls from the chains of illusory perfection.

Thus, through these painful memories, a portrait emerges of humanity searching for meaning and truth. Our common destiny may lie in our ability to recognise and overcome this great misunderstanding and embrace the freedom born of disillusionment. It is in this endeavour that we find hope for rebirth, where the light of lucidity dispels the shadows of doubt and paves the way for a return to authenticity.

Paradoxical Serenity: Freedom Born of Disillusionment

Serenity, that enigmatic notion at the heart of our tormented existence, seems paradoxical and elusive. Indeed, it often emerges from the ashes of disillusionment like a phoenix rising from its own torments. It is in this state of destitution that authentic freedom is revealed, liberated from the illusory chains of perfect happiness.

When our idealistic aspirations crumble under the relentless weight of reality, the dissonance between our expectations and the intrinsic nature of existence shakes our certainties. However, within this inner turmoil, an unexpected freedom germinates, stripped of the deceptive trappings of perfection. It is in the lucid acceptance of humanity's imperfect nature that true serenity flourishes.

The supreme pleasure of this paradoxical serenity lies in the poignant quest for meaning within absurdity itself. This is not passive resignation, but a silent rebellion against illusory mirages, revealing an unexpected spiritual freedom. This freedom gives rise to a unique force, fuelled by confrontation with disillusionment, enabling individuals to delve into the depths of their being and transcend the boundaries of the human condition.

Thus, paradoxical serenity becomes the fertile ground in which the soul's authenticity flourishes. It offers the lost spirit an inner compass, guiding its steps firmly across the bleak plain of existence. Through this intimate liberation, the individual frees themselves from vain aspirations, embracing the subtle beauty of imperfection and revealing unexpected fulfilment — a world away from fleeting chimeras.

In this inner struggle, disillusionment becomes the crucible of salvation in which a new pact with life is forged. It is in the silence of bitter renunciations that the vibrant echo of regained freedom resounds. Paradoxical serenity is the subtle alchemy that transforms the debris of broken dreams into a treasure chest, bearing witness to humanity's infinite quest for elusive harmony.

A Subtle Song in the Desert of Inner Exile

The heavy silence of inner exile conceals a secret melody: faint, yet unalterable; whispered by the stray breaths of wandering souls. In this desert of intimate solitude, the symphony of the inner self resonates with disconcerting subtlety, defying simplistic interpretation. It is in this redemptive aridity that a voice awakens: fragile yet imperious; carried by the capricious winds of existence.

The uncertain footsteps on the burning sand of inner isolation are not in vain; each footprint testifies to the desperate search for an elusive oasis — an illusory refuge to which the lost soul clings in search of meaning. This gentle lament then transcends the void, becoming a labyrinthine musical composition where the dissonances of the soul mingle with the nebulous harmonies of the mind.

Thus, wandering becomes the paradoxical cradle of an

unexpected freedom, freeing us from the chains forged by a superficial reality. Far from the turmoil of the outside world, inner exile creates an intimate amphitheatre in which the ineffable whispers of the soul are expressed. These distant echoes and subtle songs resonate in the depths of the self, revealing truths buried beneath layers of imposed conventions.

On this arduous journey, the fragile brilliance of tacit wisdom emerges from the depths of inner exile, transcendent and immutable. The seemingly endless desert takes on a new meaning, becoming a blank canvas on which the contours of an unexpected self are sketched, free from the illusions of the outside world. It becomes the sacred theatre of an archaic rebirth and an intimate communion with the unspeakable mysteries of existence.

This is the subtle song heard in the desert of inner exile: the ethereal melody of resilience where each note carries the unalterable echo of the relentless quest for truth. The inner nomad thus walks not towards a specific destination, but towards the elusive horizon of their own essence. They are accompanied in secret by this little-known song, a faithful guardian of the treasures buried in the shifting sands of the tormented soul.

2

Science Conquers Mystery

Between Fact and Fantasy

The Universe Unveiled: In Search of its Secrets

In the theatre of the stars, humanity has long gazed at the splendour of the heavens with a mixture of fascination and perplexity. Once upon a time, the universe jealously guarded its mysteries, toying with the daring minds that dared to penetrate them. However, through the prism of science, we have begun the inexorable unveiling of cosmic phenomena — a perpetual quest to unravel the intricacies of space and time. Like intrepid explorers, we brave the unknown, embracing both the growing clarity and the persistent darkness of the universe. Celestial discoveries have revealed the enigmatic structures of galaxies, the harmonious movements of planets, and the distant echoes of supernovae. Each meticulous observation and each theory developed brings us closer to the prodigious frontiers of existence, awakening an insatiable thirst for knowledge within us. Yet, despite our growing knowledge, the secrets of the universe remain shrouded in veils of immemorial mystery that defy our understanding. Quantum physics, the study of black holes and the intimate mechanisms of matter and antimatter, represent dark frontiers where reason encounters its own limitations. Thus, the infinitely small eludes our perceptions, hidden behind the mists of probability and defying the apparent order of things. Despite our cognitive arrogance, our attempts to strip the universe of its mystical trappings highlight our humble condition as bold explorers, caught between light and darkness, fleeting certainties and indecipherable abysses.

Reason in Exile: When the Invisible Outruns Calculation

When we peer into the depths of the universe, reason itself appears to be exiled. Can we truly impose boundaries on the invisible and quantify the immeasurable? The power of logic dissolves in the abysmal darkness of that which cannot be measured. This is where the beauty of the irrational emerges, defying the narrow limits of calculation. In this exile of reason, the imperceptible becomes new territory to explore.

Efforts to map the invisible are a delicate balancing act on the edge of uncertainty. Attempting to capture the infinite within the confines of mathematical formulas and physical theories also necessitates humility in the face of the elusive. Is confronting these boundaries of the explainable a defeat of reason or a victory of the human spirit? Pioneering scientists find themselves torn between the quest for rigour and the mystical allure of intellectual digression.

In this exile, we are led to embrace darkness with a certain joy. It is in the shadows of certainty that the imagination is set free, creating a space where the poetry of thought nourishes the mind. Where reason hesitates, the daring wanderings of the mind become the focus of a new understanding. In this dance between light and darkness, the intangible takes shape, defying all attempts to contain it.

At the heart of this confrontation, the human spirit awakens to a symbiotic dance with mystery. Far from any calculable finitude, it forges a path towards an alternative reality. This is a reality where rational limitations do not dictate the way forward, but rather provide fertile ground for new paradigms to emerge. This exile from reason then becomes the cradle of unexpected creativity, giving birth to ideas that were once considered impossible.

Thus, on this journey towards the invisible, reason discovers its true strength lies in its inability to explain everything. It discovers an unexpected power in its exile, defying the constraints of the known world to embrace the unknown. Perhaps it is here, in this apparent withdrawal, that reason reaches its zenith, adorned with the dazzling colours of the ineffable.

Beyond Convention: Imagination and Discovery

The history of science is littered with moments when imagination has played a pivotal role in significant discoveries. Often, it is intuition and creativity that guide researchers towards unexpected truths. In this quest for knowledge, scientists find themselves confronted with the unknown — that boundless territory where imagination reigns supreme.

Rather than being contrary to rational thought, imagination is in fact its indispensable companion. It is imagination that enables us to conceive bold hypotheses, envisage new paradigms, and expand the boundaries of knowledge. Thus, it is through the power of imagination that enlightened minds have transcended established conventions and ventured into uncharted territory.

Scientific discovery cannot be limited to a rigid, methodical approach. Above all, it requires an openness to the unexpected and a willingness to marvel and explore. Rather than hindering scientific rigour, imagination enables us to envisage parallel universes, unexpected dimensions and revolutionary theories.

Sometimes, imagination is crucial in solving seemingly unsolvable problems. It is in these moments of deadlock that the inventive mind thrives, exploring new avenues and breaking with conventional patterns to arrive at innovative solutions. Great scientific revolutions have often emerged from imaginative daring, challenging conventional wisdom and shaping a new horizon of knowledge.

Ultimately, imagination and scientific discovery are inseparable. They intertwine to transcend the barriers of existing knowledge and give rise to revolutionary advances. This alliance between imagination and reason paves the way for scientific progress, offering curious minds the chance to explore the universe through a new lens, filled with fascination and wonder.

From the Original Enigma to Modern Revolutions

In a world where knowledge emerges from the darkness of ignorance, it is essential to immerse oneself in the mysterious depths of the original enigma. Since taking our first steps on this secret-rich land, humanity has bowed before the great mysteries of the universe, desperately seeking to pierce the veil of time and discover the ultimate truth. Ancient civilisations left behind mythical tales and esoteric legends, highlighting the endless quest for knowledge. Creation stories, cosmic deities, and prophetic ballads have formed the basis for exploring the primordial mysteries that have haunted the human mind since the beginning of time. Ancient philosophers attempted to decode the essence of existence, but were confronted with an impenetrable wall of origins. Schools of thought have succeeded one another, shaping and transforming the perception of the supreme truth without ever fully revealing it.

Science as a Contemporary Myth: The Illusion of Total Knowledge

Modern scholars often find themselves prey to the seductive illusion of total knowledge. Following the scientific revolutions of previous centuries, contemporary humans tend to believe in the omnipotence of rational knowledge. However, this blind confidence can mask the complex and elusive nature of the universe. This scientific arrogance, rooted in an unwavering belief in the human capacity to comprehend the vast and the minuscule, resembles a modern-day myth.

The pursuit of absolute knowledge is inevitably confronted with the darkness that surrounds our certainties.

Like fleeting fireflies in the cosmic darkness, our theories and discoveries momentarily illuminate our path but never dispel the thick fog of mystery. Our tendency to hide our ignorance behind a deceptive veil of total understanding is the very essence of the illusion of knowledge.

Yet this illusion does not merely lead us astray. It also inspires an unquenchable quest for truth and an insatiable thirst for light in the heart of darkness. It is in this confrontation between shadow and light that a fertile dialectic is born, capable of taking science to new horizons. Recognising our ignorance is thus not a weakness, but a springboard for deeper exploration of the unknown.

It is in this space between darkness and light that the struggle of the Enlightenment plays out. It is a relentless struggle made up of doubts and questioning, where each advance marks not the end of the quest, but its infinite renewal. Thus, rather than being a fixed code of absolute truths, science reveals itself to be a perpetual dialogue with the inexplicable and a continuous drive towards exploring new mysteries. In this endless quest, humanity rediscovers its modest place in an infinitely vast and complex universe where each answer opens the way to new, fascinating questions. Ultimately, science — this contemporary myth — becomes a permanent invitation to humility and to marvelling at the immensity of reality.

Between Darkness and Light: The Struggle of the Enlightenment

Human passions merge in an eternal dance between

darkness and light — a relentless struggle that lays bare the fragility of our condition. Like a mask removed to reveal the face of the soul, science has two facets: one dazzling, the other enigmatic. It is in this theatre of contradictions that the very essence of human knowledge unfolds. Darkness, like a moonless night, engulfs our certainties, confronts us with our limitations and reminds us of the limitations of our knowledge. Yet light, like the sun rising over an infinite horizon, inspires hope and a desire to reach new illuminated horizons. Thus, we are caught up in this elusive duality, oscillating between the vulnerability of our ignorance and the sparkling promise of knowledge. The perpetual struggle between light and darkness reveals the mysteries of the cosmos and the dark recesses of the human soul. It is in this fragile balance that the contours of our perpetual quest are drawn: a quest that transcends the limits of understanding and bears the indelible mark of our humanity. It is at the crossroads of darkness and light that the deepest questions flourish, forging the timeless epic of discovery. As we explore the unknown, clarity beckons us like an elusive muse, providing an endless source of inspiration and achievement. However, we cannot ignore the other side of the coin: behind every flash of light lurks the disturbing shadow of uncertainty. Thus, the struggle between light and darkness shapes our destiny, intermittently illuminating the steep paths of understanding but also tinging our certainties with dark hues. In this relentless struggle, humanity shapes its relationship with the world, imprinting it with its ever-evolving consciousness. Therefore, it is up to us to embrace this fascinating duality with humility and navigate the troubled waters of the unknown courageously, while embracing the fleeting beauty of the revelations scattered along our path.

When Theories Collapse: The Sinking of Belief

Theories stand before us like impregnable fortresses, challenging our understanding and our ability to grasp the world's intrinsic complexity. Like an indomitable tide, these concepts, built on fragile foundations, sometimes succumb to the shipwreck of certainty, leaving an ocean of uncertainty in their wake. It is in these tumultuous moments, when the pillars on which our understanding rests seem to falter, that the true test of thought lies. For it is in these troubled waters that we must navigate without a compass, leaving the familiar shores of knowledge behind to explore the uncertain horizons of questioning. However, beyond this intellectual crisis lies a deeper revelation: the awareness of our own ignorance and the humble recognition of the cosmos's unfathomable grandeur. Theories may collapse and certainties may waver, but it is in this void of uncertainty that the seeds of discovery germinate, ready to blossom in the fertile soil of the curious mind. It is there, in the abyss of the unknown, that the human mind can renew its view of the world, free itself from preconceptions, and let bold thinking flourish. Thus, when theories collapse, we are not condemned to darkness, but invited to embrace the dawning light of a universe in perpetual redefinition.

Humans Facing the Infinite: Memory and the Ephemeral

The human condition is perpetually confronted with the immensity of the universe, yet it is also marked by an inevitable transience. Endowed with memory, humans constantly measure themselves against the ever-expanding infinite, feeling a growing gap between what is retained and what is irretrievably lost every day. Memory is both a faithful companion and a cruel accomplice, reminding us of the fleeting nature of our existence. Moments of contemplation reveal the duality of this experience: the confrontation between memories that seem eternal and the transience of all earthly things. This paradoxical state of being gives rise to an insatiable quest to capture fleeting moments and make them immortal in the fragile web of memory. However, as soon as humans attempt to grasp the past, they cannot help but feel life's incessant metamorphosis, like a perpetual wave sweeping away everything once acquired. Thus, humans oscillate between these two extremes like tightrope walkers on the thread of time, seeking to make the elusive immortal and engrave the fragility of their existence in stone. This perpetual struggle against the ephemeral transcends all boundaries and eras because it is a universal drama that inhabits every consciousness. So, how can we embrace the ubiquity of infinity in the fleeting dance of our lives? How can we tame memory, which is both preservative and destructive, in order to fully savour the flashes of infinity that sparkle in our everyday lives? These questions resonate within each of us, inviting us to delve into the depths of the human soul — that unfathomable realm where the infinitely large and the infinitely small converge — and discover the hidden treasures of our existence.

Reflections of the Unknown: The Beauty of an Unfathomable Everyday Life

The unknown, our eternal companion, envelops us like an elusive mist, brushing against our thoughts as the days go by. In this dark abyss of ignorance, a mysterious beauty shimmers in the dim light of our profound questions. Each step we take on the path of knowledge seems to illuminate a fragment of this unexplored territory, revealing dazzling reflections and unexpected nuances. Our minds delight in wandering through the twists and turns of this unfathomable everyday life, captivated by its infinite richness.

Contemplating the perpetual dance of the known and the unknown is fascinating: a celestial symphony in which familiar notes compete with unexpected melodies. At every moment, we witness the eternal encounter between the familiar and the strange: an enchanting alchemy that unveils the hidden splendour of our reality. The enigmatic becomes the focus of our meditations, illuminating our paths with a dazzling light that reveals the essence of our existence.

Like intrepid explorers, we courageously venture into

the depths of the unknown, discovering unexpected treasures and elusive truths that defy definition. The beauty of this quest lies in the constant wonder it inspires, like an inexhaustible source of amazement and admiration. Each fragment of the unknown holds the promise of revelation and invites us to contemplate the universe with fresh eyes and reconsider our place in this vast cosmic ballet.

At the heart of this contemplation lies an aura of humility and a deep recognition of our insignificance in the face of infinity. We are lulled by the constant oscillation between what we know and what we do not, carried along by the tumultuous flow of discovery and revelation. The beauty of the unfathomable everyday unfolds before our astonished eyes like a living fresco, whose ephemeral hues bathe our existence in lush, intoxicating poetry.

Thus, in this enchanting embrace with the unknown, we find a refuge where the wisdom of the inquisitive and the passion of the tireless seeker intermingle. In every enigma and mystery, we glimpse the essence of this elusive beauty that permeates our being. Walking the steep paths of the unknown means embracing the adventure of existence and appreciating the fascinating intersection of light and darkness, knowledge and ignorance.

At the Frontiers of the Forbidden: Transcending Knowledge

Eager for knowledge, human beings have always sought to expand the boundaries of their understanding. Like a shifting frontier, this perpetual quest for comprehension continues to challenge us. But what lies beyond the limits imagined by our curious minds? Venturing into risky territory at the edge of the unknown, humans push back the barriers erected by fear and ignorance, braving the forbidden. Scientific research, like artistic exploration, is a constant struggle against our own limitations.

Like mountaineers searching for unclimbed peaks, we boldly climb the vertiginous slopes of mystery. The mind seeks to unveil the seemingly unattainable and desires to transcend the boundaries of the known, immersing itself in the vast expanse of the unknown. Bold discoveries bear witness to this relentless pursuit of knowledge, in which humanity, like Prometheus defying the gods, wrestles nature's best-kept secrets from its grasp.

However, this journey towards the forbidden is not without danger. Crossing the ethereal boundaries of knowledge brings us face to face with the shadow — that dark side which even enlightened knowledge can never entirely dispel. Like Icarus flying too close to the sun, humanity fears a precipitous descent into the abyss of ignorance. The temptation to transcend the limits of what is known can lead to confusion and the embrace of ignorance under the guise of apparent knowledge.

Nevertheless, beyond these risks lies the greatness of the human aspiration to transcend established boundaries. This insatiable desire to transcend the known is the driving force behind our evolution. In every scientific advance or artistic achievement, humanity stands before the forbidden, ready to cross the threshold into the unknown. It is in this act of bravery that we find the essence of our humanity: the ability to confront darkness and move forward despite the shadows that threaten to overwhelm us.

Ultimately, at the frontiers of the forbidden, the pursuit of knowledge is a perilous adventure and an act of faith in our infinite potential. By crossing these fragile boundaries, humanity reveals its ability to rise above itself and embrace the distant horizon of knowledge. This eternal quest echoes our wildest dreams and promises a future written in the stars of infinity.

3

The Siren Song of Art

Emotions and imagination run free

At the Origins of Dreams: The Birth of Art

Since the dawn of time, humans have sought to capture the mysteries of life through art. From the cave paintings of early hunter-gatherers to modern masterpieces, each era has seen unique expressions of the human spirit emerge. Cave art, for instance, testifies to our ancestors' quest for meaning and their contemplation of nature. The origins of sculpture and pottery in ancient civilisations demonstrate the close connection between matter and creative imagination. The first Greek tragedies expressed the deepest human emotions, while medieval art, imbued with the sacred, offered a reflection of transcendence and faith. The Renaissance saw a renewed exploration of humanity through painting and sculpture, while 19th-century Romanticism produced works that evoke emotion and introspection. Every artistic movement has contributed to our ever-evolving perception of the world. From music to literature and dance to theatre, each art form has exerted a transformative influence on the human mind, offering a timeless journey through emotion and imagination. Therefore, a work of art is not merely the product of technical skill; it is the result of an encounter between the material world and the artist's inner world. It reflects the essence of humanity and opens doors to unimagined universes.

Emotions in Motion: The Dance of the Imagination

Art is the eternal song of emotions in transformation, a spellbinding dance that reflects the deepest impulses of the soul. Like a pas de deux between the artist and their inspiration, the dance of the imagination transcends the limits of the tangible to embrace the invisible. Every brushstroke, every musical note and every word carves out space and time, creating a symphony of inexpressible beauty. Art's vital force springs from the source of our emotions, winding its way through the labyrinth of the human mind to reveal its pure essence more clearly. Emotion thus becomes the master of ceremonies, the supreme orchestrator of this aesthetic dance. In this swirling momentum, the mind surrenders to the mysteries of the imagination, unhindered and unconstrained. The artist's passion blends irresistibly and fiercely with their audience's infinite sensitivity, transforming the work into an emotional ballet where feelings harmonise and clash in a timeless dialogue. The dance of the imagination spreads its dreamlike wings, carrying the soul to unimaginable horizons. Within it lies the promise of an exhilarating and dizzying journey into the unknown, where the very essence of creation is woven. The artist's emotional quest becomes a unique odyssey — an exploration of the meanderings of the human soul. Everyone becomes their own choreographer, creating an emotional tapestry where colours, sounds, and words blend into an inextricable symphony. This dance offers a gateway to the ultimate understanding of our humanity, elevating our daily lives and offering them up to the grandeur of the universe. Thus, emotion ignites art, awakening numbed consciousnesses and giving birth to this theatre of the soul, where each spectator can find their place and resonate with the rhythm of this ancestral dance.

Under the Sky of the Muses: The Artist and Their Inner Light

Like a receptive filter of the world's mysteries, the artist finds themselves immersed in an ocean of emotions, sailing on waves of inspiration radiated by immortal muses. On this inner journey, they explore the depths of their soul, searching for the inner light that illuminates their creative path. Like an alchemist of the mind, they celebrate the sacred union of their innermost thoughts and the whispers of the universe. It is under the sky of the Muses that they become the conduit through which transcendental artistic grace unfolds. Every emotion and thought is transformed into a palette of dazzling colours, creating a work charged with meaning and pulsating with life. This inner light, reflecting distant stars and inner turmoil, is a beacon in the darkness of existence, guiding the artist towards unimagined horizons. Indeed, like a North Star, it gives them the courage to confront the unfathomable mystery of artistic creation. It is in the silence that permeates the depths of their being that artists distil the elixir of their genius, in contemplation and listening to the celestial harmony that reveals its subtle nuances. Sometimes, this elixir is born from the shadows as the mind plunges into the unfathomable depths of the human soul and finds the gushing source of inspiration in its darkness. Thus, the artist, vesperal and winged, draws their divine strength from these moments of calm, where each breath becomes a silent stanza in the great cosmic poem of existence. Their inner light is not fixed, but in constant metamorphosis, like the aurora borealis caressing the firmament of human aspirations. It guides their movements, shapes their gestures and distils the primordial energy of all authentic creation. Under the sky of the Muses, the artist finds their timeless voice — an endless path to infinity.

The Vibration of Silence: Between Creation and Contemplation

In silence, the artist finds refuge. It is in this sacred void that the deepest ideas and purest emotions take root. Creation is born from this inner vibration, from silent meditation where the soul unites with the universe. In this timeless space, the artist can hear the gentle, melancholy voice of inspiration. They contemplate the void and discover an unexpected fullness within it. Here, works are born that transcend the limits of reality to touch infinite possibilities. Contemplation then becomes the keystone of artistic expression because it is by losing oneself in silence that one truly finds oneself. Each musical note, each brushstroke, is a tribute to silence, a dialogue with the inexpressible. It is in these moments of grace, between inhalation and exhalation, that the artist finds balance and creativity unfolds freely, where doubts fade away. Thus, silence becomes the fertile ground of the imagination and the inexhaustible source that feeds all artistic expression. In this communion with silence, the artist becomes one with the essence of the world, and their creations become vibrant reflections of this magical fusion. This perpetual state of meditation offers a fresh perspective on reality, enabling the artist to explore uncharted territories and expand the boundaries of the visible to gain a deeper understanding of the invisible. The vibration of silence thus resonates in each of the artist's works, inviting viewers to dive into an ocean of emotions and dreams, where creation and contemplation become one.

The Reflection of an Evanescent World: When Art Reveals Reality

At the heart of every work of art lies a reflection of an evanescent world: a reality that is sometimes veiled and invites the viewer's soul to immerse itself in the depths of creative expression. Like a masterful mirror, art transcends the limits of the tangible to capture the essence of human emotions and universal truths. Through every brushstroke, musical note or written word, the artist sketches the fragile yet intense contours of our existence, illuminating the dark corners of our consciousness. By revealing reality in a new light, art invites us to contemplate the elusive beauty emanating from our human condition. Each masterpiece exudes a unique poetry, revealing the disillusions and hopes that shape our daily lives. In this way, art has the power to awaken our consciousness, challenge our certainties, and reveal the paradoxes inherent in our nature. Through the subtlety of its nuances and the power of its contrasts, art offers us a prism through which we can grasp the complexity of life and savour its richness. Indeed, every artistic creation invites us to explore the uncertain territories of emotion, to engage in dialogue with our dark and light sides and to tread the fine line between illusion and reality. Thus, art becomes a vehicle for revealing subtle truths that transcend objectivity to embrace intuition and emotion. Moments of contemplation, immersed in the depths of a work of art, offer a renewed vision of our surroundings, tinged with mystery and sublimity. Art orchestrates an endless symphony in which harmony and chaos intertwine, weaving invisible threads that connect us to the world and to ourselves. When art reveals reality, it unleashes a creative force within us, heightening our sensitivity and inviting us to embrace the essence of our humanity.

Storms of Colour: Harmony and Chaos in Symphony

Colours are the whispers of the immortal soul that weave the ephemeral fabric of our days. They unfold in a choreography that is by turns violent and gentle, embracing the canvas and awakening our senses, which have become numbed by the greyness of everyday life.

Bright hues intertwine in an infinite dance, where harmony and discord, light and darkness mingle like silent cries. Each brushstroke is an unexpected storm; each nuance, a secret revealed. In this pictorial tumult, the painting reveals its own music: an invisible wave that stirs and caresses our deepest emotions. Thus, the chaos of colours is adorned with a subtle symphony, reviving the forgotten chords of our humanity within us. Like clandestine conductors, the masters of art orchestrate this ocean of hues, embracing human passions in a waltz of pigments. Swaying between intoxication and contemplation, viewers become actors, plunging into the heart of this tumultuous symphony. Each work becomes an emotional kaleidoscope, exhaling the scents of life with every breath. In the face of this colourful turmoil, the soul transcends appearances and awakens to a new reality imbued with the enchanting allure of marvellous canvases. Harmony is no longer a quest, but an obvious truth emerging from the mysteries of non-conformity. Meanwhile, chaos becomes the raw material for unlimited creation, far from being a threat. Finally, in this divine splendour, a gateway to a universe of unprecedented sensations emerges, where each storm of colours sings an intoxicating hymn to artistic freedom.

The Alchemy of Words: Poetry as Resistance

The alchemy of words is a powerful force that transcends the ordinary to become a weapon of resistance, echoing the voices of poets who defy their relentless fate. Like a master craftsman chiselling the ephemeral from raw stone, the poet wages a silent war, building ramparts of verse against oppression and banality. In this alchemy, words are transformed into catalysts of raw emotion, revealing the depths of existence to those with sensitive souls. Poetry becomes a refuge and a sanctuary where the thirsty can quench their thirst, where flashes of truth sparkle in the heart of darkness. Each alexandrine is a sharp blade cutting through stale air, and each rhyme ignites dormant hearts, rekindling the flame of hope.

Like a wildflower growing in the cracks of asphalt, poetry refuses to bow to the injustices of the world. It stands like a beacon in the night of numbed consciences, illuminating the paths to freedom and humanity. In this universe of resistance, verses become stelae inscribed with the pain and hope of a whole people, where evils are transformed into elixirs of revolt and the desire for change. Thus, the beauty of poetry becomes a formidable weapon, piercing indifference and awakening dormant consciences. It invites everyone to contemplate the universe with fresh eyes and embrace reality with renewed boldness.

Nourished by the ink of rebellious artists, poetry becomes an oasis in the desert of mediocrity, offering inspiration and hope to those searching for meaning. Through its enchanting metaphors and incandescent verses, poetry overturns established codes, shatters rigid dogmas, and celebrates the inexpressible. Transforming torment into oracles, it guides humanity towards a bright, unconstrained and misunderstanding-free future. Thus, through the alchemy of words, poetry becomes a bastion of resistance, urging everyone to raise their voices in an unalterable symphony of love, brotherhood, and freedom.

From Confinement to Awakening: Art as Liberator

What is the capacity of art to free the mind and soul from the constraints and torments of everyday life?

Art has always been seen as a breath of fresh air, transcending the limitations imposed by society. Throughout the centuries, artists have used their creativity to challenge established norms and conventions, opening doors to new horizons of thought and expression.

Contemplating a work of art, be it a vibrant painting, a majestic sculpture or a captivating poem, invites us to broaden our perspectives and explore inner worlds we had not previously considered. This transcendence from tangible reality to a deeper, more enriching one lies at the heart of art's liberating power.

Take literature, for example. Great writers often challenge the barriers of conventional thinking by offering narratives that question social and moral dogmas.

Through their words, these writers capture the essence of humanity, breaking the chains of conformity and inspiring readers to free themselves from the limitations imposed by society.

Similarly, artistic movements have often been acts of resistance against cultural and political oppression. By experimenting with new forms of expression, artists have transcended the boundaries of social acceptance, challenging the rigid and restrictive structures that sought to imprison them.

In this way, art becomes a tool for liberation, enabling individuals to turn away from conformity and connect with deeper truths. Reflection and creation offer a path to awakening, allowing individuals to transcend the limits of ordinary existence and reach a higher level of understanding and sensitivity.

In short, art liberates the soul, offering humanity the possibility of rising above the constraints of the human condition.

The Sirens of Modernity: Back and Forth

At a time when the world is embracing the nuances of modernity, art finds itself ensnared by the allure of modernity while attempting to escape its captivating charm. Like these elusive mythological creatures, modernity offers an infinite space for artistic exploration but also poses a formidable trap for the creative soul. The tension between tradition and the avant-garde, between bold innovation and preserving roots, is palpable in the contemporary artistic universe. Like seasoned sailors, artists skilfully navigate the turbulent waves of cultural evolution, seeking to strike a delicate balance between reinvention and continuity.

Modernity's allure lies in its promise of renewal, propelling creators into uncharted territory where audacity and transgression reign supreme. However, this temptation invariably meets resistance in the form of a thousand-year-old artistic heritage that remains solid in the face of novelty. The sirens of modernity call for a complete break with tradition, but artists, faithful to their vocation, know that true treasures are sometimes buried at the heart of history itself.

This constant interplay between fascination with the new and respect for the past fosters a rich dialogue within contemporary art. Works are thus animated by dynamic tensions, blending the echoes of old masters with the frenzy of visionary creation. This perpetual confrontation between the old and the new, between permanence and transformation, gives rise to a unique aesthetic imbued with the contradictions of its time. Through this incessant movement, art transcends temporal boundaries, forging unexpected links between eras and offering a kaleidoscopic vision of humanity.

Ultimately, the sirens of modernity urge artists to embrace the ambiguity of their era and follow the tumultuous currents of innovation, all the while remaining anchored in the richness of cultural heritage. It is in this oscillation between the ephemeral and the eternal that artistic fertility germinates, infusing each work with unparalleled depth. Thus, the interplay between seductive modernity and saving tradition endows contemporary art with an unquenchable vitality — a living symbol of the perpetual quest for meaning and beauty in a world that is constantly changing.

Poetics of Horizons: Towards an Artistic Infinity

In an ocean of reflections, the artist ventures towards distant horizons of infinite expression. Like a lone sailor on a ship of thoughts, they navigate the turbulent waves of creation, tirelessly searching for a magic formula to capture the evanescence of time and shed light on the mystery of human existence. The sirens of modernity sing their haunting melody, urging the artist to navigate the pitfalls of tradition and the turbulent currents of novelty. Beyond these challenges, the artist finds solace in the poetry of horizons, where sky and earth converge to create an infinite canvas. Every colour, shape and line on this canvas is an invitation to explore the boundless creativity of the human spirit. In this endless quest, the artist becomes a conduit for emotions and a privileged observer of the human soul, revealing the infinite richness of inner landscapes with subtle touches. Thus, artistic infinity reveals itself as a magical territory where imagination is free to unfold without constraint, where beauty is expressed in myriad forms and the universal language of art transcends borders to forge deep connections between souls. It is in this frenzied dance of ideas and forms that the artist discovers their true essence, merging with the eternal work they have created, forever moved and forever moving. Thus, the horizons stretch out to infinity, inviting the artist and the viewer to embark on an endless journey to the heart of emotions, to the heart of the universe and to the heart of the self.

4

When Faith Triggers Divine Laughter

Is the House of God a Refuge?

The Echo of the Sacred in a Disenchanted World

Modern existence, steeped in scientific progress and rationality, calls into question the presence of the sacred in everyday human life. However, even in this world devoid of mystery, echoes of the sacred still resonate. The progressive secularisation of society has not stifled the age-old need for transcendence. Faced with widespread disenchantment, the human soul is torn between an insatiable thirst for meaning and a reality stripped of all spirituality. In this universe seemingly devoid of miracles, the echo of the sacred nevertheless persists, lurking amid the rational quirks of the modern world. It can be found in the rustling of autumn leaves, the unfathomable depths of a child's eyes and the intoxicating melody of a piano played by an artist possessed by divine inspiration. Despite the dominance of reason, the echo of the sacred spreads its invisible wings over humanity, which is hungry for transcendence. From silent contemplation of the starry sky to the soothing sounds of nature, the sacred finds refuge in the most unexpected places in our disenchanted world. Thus, even in this rational age, every breath of wind carries with it a sigh of transcendence and every ray of sunshine caresses the soul in search of the absolute. Far from being confined to the depths of history, the sacred has been transformed into an intimate experience that permeates the contemporary consciousness like a phoenix rising from the ashes in the rationalist universe.

God in the Image of Our Laughter: A Lucid Reflection

The very notion of the divine is inscribed in the lines of your face when you smile, like a subtle reminder that is both dazzling and elusive. The essence of divine splendour is interwoven with the contours of your laughter, bursting forth like distant stars in the infinite firmament. As we examine this fleeting glimmer on the horizon of our understanding, we are confronted with an enigma in which transcendence is intimately intertwined with the ephemeral. The striking paradox of modern stagnation is the relentless quest for meaning slowly being eroded by repeated blows of absurdity. Our laughter seems to resonate like a mesmerising prism, shaking certainties and overturning immutable precepts. In this grandiose spectacle of life, God adorns himself with the shimmering hues of humour and assumes the radiant mask of lucidity. The illusion of divine omnipotence fades before this enlightened laughter, revealing profound humanity and indispensable fragility. In this lucid reflection, a universe emerges in which faith is anchored in the arid soil of existence and takes root in the heart of our paradoxes and questions. Transcendence then takes on the fabric of consciousness itself, freeing itself from dogma to embrace the disruptive vitality of our heartfelt laughter. Beyond the tumult of rigid interpretations, a vision emerges that celebrates the delicate intimacy between faith and questioning, forging unexpected connections between the sacred and the profane. This God, imbued with the flavour of our laughter, accompanies us on our chaotic journey through life, transforming fixed boundaries into endless horizons of discovery. Thus, the light breeze of divine laughter blows across the arid expanses of our shaken certainties, breathing new life into our relationship with transcendence. In this dazzling blossoming lies the seed of essential freedom, urging us to listen to the majestically dissonant symphony of faith and questioning.

The Paradox of the Modern Believer: Between Faith and Questioning

In the modern world, believers face a striking paradox. On the one hand, they remain anchored in their faith, which is an inexhaustible source of comfort and hope. But, on the other hand, these same believers are plagued by incessant questioning that shakes the very foundations of their convictions. This duality — or, rather, this tragic duality — is strikingly evident in the psyche of contemporary believers.

It is impossible to ignore the fact that our era is characterised by an insatiable thirst for truth and rationality. Scientific and technological advances confront us with challenges that stimulate our critical thinking and fuel our doubts. Steeped in this sceptical atmosphere, the modern believer finds himself immersed in a perpetual quest for meaning.

Alongside the influence of rational thinking, believers must also contend with a social and cultural environment characterised by diverse beliefs, and even a rejection of all forms of spirituality. Confronted with such plurality, they are compelled to test their faith against difference, leading to perpetual questioning of their convictions.

This inner tension between faith and questioning can lead modern believers to a state of deep perplexity and even an identity crisis. They must juggle the certainty of their faith with the urgent need to question its foundations relentlessly. In this precarious balance, faith becomes a desperate quest for understanding, self-transcendence, and openness to otherness.

Furthermore, modern believers experience moments of uncertainty when they are tempted to succumb to the simplicity of dogma rather than embrace the inherent uncertainty of any spiritual journey. Fundamentally inhabited by doubt, their spiritual journey is reminiscent of Socrates', who acknowledged his wisdom in the face of his own ignorance. Thus, the path of the modern believer is akin to a tortuous pilgrimage, punctuated by existential questions and insidious quests. This arduous path is strewn with obstacles, but it also offers unexpected inner riches, forged in the crucible of reflection and introspection.

Rituals and Silence: When the Temple Becomes a Theatre Stage

Are these endlessly repeated gestures the place where the essence of the sacred is revealed, or are they simply meaningless theatrical performances devoid of true communion? This is the central question that animates our reflections. Silence, like a gently rising curtain, envelops these sacred scenes in its deep folds. Each movement then becomes a subtle dance between the visible and the invisible, the observable and the transcendent. In this timeless theatre, actors and spectators converge to offer a performance in which every repeated gesture has the potential to reveal something new. Thus, silence becomes the vehicle of the sacred itself, inviting everyone to free themselves from the banality of everyday life and ascend to higher realms. However, when the sacred overflows its boundaries, carrying with it an excess of fervour, are we not in danger of slipping inexorably into superstition? When exceeded, the limits of the sacred are fragile boundaries where the shadow of doubt looms large in the face of ritual excess. Points of reference fade, giving way to confusion and questions. Therefore, everything hinges on this fine line between fervour and superstition. This sacred theatre, where ancestral gestures and modern clamour intermingle, remains a space of perpetual questioning and fertile ground for reflection. Silence is alternately exalted and sullied in this spiritual theatre, like a pendulum swinging between authentic contemplation and agonising mutism. Our gaze is drawn into this interplay of light and darkness, prompting increasingly acute questions about the significance of these rituals and the weight of silence in this sacred setting. Through these reflections, the many facets of the temple emerge, resonating with a spirituality that is both ancient and ever-renewed, where the sacred pulses to the rhythm of modernity's changing lights.

Fervour or Superstition: The Limits of the Overflowing Sacred

The fervour that animates believers during religious practices can be a powerful catalyst for spiritual communion. However, these outbursts of deep piety can sometimes tip over into superstition. When devotion turns into rituals designed to appease existential anxieties, we must question the limits of the sacred.

The line between religious fervour and superstition is often blurred. Faith can offer comfort and direction, but when beliefs become intertwined with ritualised practices that are disconnected from their essence, the distinction between true spirituality and superstition becomes unclear. Such rituals can then prove to be poor substitutes for a genuine connection with the divine, condemning the faithful to a perpetual cycle of fear and doubt.

Secular traditions and ancestral rites can also obscure an authentic understanding of faith. In the confusion of symbols and gestures revered for generations, the original meaning can be lost, leaving only a mechanical practice devoid of real significance. The distinction between the sacred and the profane becomes blurred, creating a belief structure built on fragile foundations.

However, humans are capable not only of being superstitious, but also of elevating their beliefs to true spirituality. Discernment, introspection, and inner reflection are key to distinguishing fervent religious practices from superstition. Only those who can transcend external forms to reach the heart of faith will be able to experience its true essence. Recognising when the sacred becomes excessive is essential to preserving the essence of spirituality. Each of us must question the notion of religious fervour and distinguish sincere faith from obscure superstition in order to move towards an enlightened inner truth.

The House of God: Spiritual Refuge or Prison of the Soul?

When we enter the House of God, a complex duality presents itself. Is it a spiritual refuge, a haven of peace amid the turmoil of earthly life? Or is it a prison of the soul, stifling emotions and questions under the weight of rigid dogma? This is a question that haunts the minds of believers and non-believers alike, as the relationship with the divine is full of paradoxes and subtle nuances. Sometimes, the House of God appears as a beacon in the night, offering an inexhaustible source of comfort and inspiration. The faithful gather there to draw the strength they need to face the trials of the outside world. Every corner of this sanctuary echoes with murmurs of prayer, haunting chants, and sacred stories that soothe tormented souls in search of serenity. Yet behind this luminous façade, a darker reality sometimes lurks. The House of God can also become a prison for thought and individuality. Rigid beliefs and dogma can force minds to conform to a single vision of spirituality, threatening freedom of thought and expression. Excessive dogmatism risks stifling the free interpretation of sacred texts and spiritual teachings, thus depriving the soul of its autonomy and its ability to explore its innermost self. The nagging question remains: does the House of God offer a safe haven for the wandering soul, or does it become an invisible prison where impulses of thought and faith are extinguished? This question transcends religious dogma, reaching the very essence of the human experience and inviting each of us to reflect on our relationship with the divine and with difference. Whether believer or not, it is up to each individual to find the key within themselves that will open the doors to their own inner House of God, transforming it into a true spiritual refuge imbued with freedom and kindness where the noblest aspirations of the human soul can flourish.

The Discordant Voices of Shared Faith: Towards Harmonisation

Amidst this tumult of intersecting beliefs, the discordant voices of the faithful resound. The sacred word, expressed according to the plurality of dogmas and interpreted according to each individual's conscience, inspires the great theatre of shared faith. The chants emanating from mosques mingle with the ringing of church bells, while the psalms of Buddhist temples rise towards the starry sky. However, beyond this diversity, subtle discord sometimes emerges, dividing souls in search of salvation. Faith then becomes a battlefield where everyone fights to impose their own interpretation of the divine.

However, within these spiritual confrontations, the breath of harmonisation also emerges. Respectful listening and serene dialogue between believers with divergent convictions are key to transcending disagreements and achieving a deeper union. It is precisely in this relentless and humble quest for mutual understanding that faith is transformed into a celestial symphony, woven from nuances and reconciling chords.

Thus, despite apparent differences, shared faith urgently calls for discordant voices to unite on a single journey towards the light. Each person, guided by their own faith, becomes a messenger and receiver in turn, contributing to this collective momentum towards a shared ideal: humanity embracing the divine mysteries with renewed understanding. By recognising the unexpected richness within the diversity of beliefs, believers open themselves up to the wonder of a shared, transcendent, and infinitely enriching experience.

The Divine in Everyday Life: A Dialogue with the Absurd

The divine, that silent interlocutor, slips into the interstices of our tormented existences, whispering its enigmatic verses to ears eager for ultimate truth. In an age where rationality has assumed the status of a demiurge, the divine is an unwelcome guest in intellectual conversations yet cannot be ignored. Emerging like an unsolvable enigma, it challenges the certainties and meaninglessness that punctuate our daily lives.

This timeless dialogue between the divine and the absurd immerses us in the heart of the human condition, where souls are adrift amid the twists and turns of fate and contingency. Every gesture and every breath carries within it the distant echo of this indomitable presence, manifest through the vicissitudes of ordinary life. Thus, the absurd becomes the eardrum on which the celestial voice resonates, transforming banality into a mystical symphony.

In this dialogue, reason falters when confronted with the insurmountable limits of the sacred. Torn between his quest for meaning and the apparent emptiness of existence, humanity is confronted with a perplexing enigma. Is the divine keystone to be found in the heart of absurdity? Or is it a cruel cosmic joke, a farce in which humans are sometimes victims and sometimes masters?

In this impromptu dialogue, questions mingle with mystical whispers, weaving the complex fabric of a transcendental experience. The daily routine is imbued with a majestic aura, elevated by the mysterious encounter between the profane and the sacred. Each step becomes a ritual dance, each smile an offering to the unknown and each silence a silent prayer.

As the absurd creeps into the folds of reality, the divine emerges as an invitation to explore the depths of the incomprehensible. This incessant dialogue quenches the soul's thirst for truth, urging us to embrace the enigma rather than dispel it. Perhaps it is in these enigmatic exchanges with the absurd that the elusive contours of the divine are revealed, guiding humanity towards unsuspected enlightenment.

In the Shadow of the Bell Towers: Stories of Omniscient Laughter

On the eroded cobblestones of the old town, the shadows of the bell towers stretch out like an ethereal canvas. In the heart of this thousand-year-old sanctuary, stories intertwine like wisps of smoke rising from incense burners. The cobbled streets, strewn with ancient secrets, echo with the murmurs of the faithful seeking solace.

Here, the essence of omniscient laughter resides in the silent dialogue between old stones and tormented spirits. Stories are woven here, each one carrying its own burden of pain that has been transformed into divine laughter. Wandering souls seem to find refuge in the simplicity of this immortal building, where the tolling of bells evokes echoes of immeasurable wisdom.

The forecourts welcome pilgrims in search of answers, their whispered prayers mingling with the ancient sighs engraved in the walls. Every vault conceals an echo of suffering and hope; every stained-glass window narrates the epic tale of humanity's search for enlightenment. In the abandoned chapels, one can hear irreverent laughter echoing, transcending earthly torments.

Thus, immersed in the benevolent shadow of the bell towers, only the melodies of ancient laughter echo, bearing witness to the absurdity of the cosmos as perceived by humanity. Far from rigid dogma, the tiny and the vast engage in a fragile dialogue, embracing the ephemeral with infinite tenderness. What unfathomable mysteries inhabit these walls? What echoes of transcendence infuse their stone substance?

Polished by the ages, each stone seems to inspire these gentle notes of joy, enveloping visitors in a timeless cocoon where light filters through the chimeras of humanity. Beneath the motionless arches, each heartbeat becomes a prayer, while cosmic laughter continues to intertwine with the absurdity of the world.

Perhaps it is here that divine laughter is revealed: intangible, yet present in every atom of this ancient building. In the shadow of the bell towers, tales of omniscient laughter whisper to souls in search of meaning, offering the comforting reassurance of a divine presence that laughs at earthly turmoil and soothes hearts thirsty for hope.

Echoes of the New Dawn: Rediscovering the Simplicity of Mystery

A new dawn is breaking over the landscape of our shaken certainties, offering a renewed perspective on the eternal mystery surrounding our existence. Rediscovering the simplicity of mystery means plunging into the heart of the inexpressible and embracing the beauty of the world in all its unexplained splendour. In a universe where the omnipresence of omniscient laughter is felt, discordant voices merge into a harmonious chorus, resonating through the sacred silence of new clarity.

Freeing ourselves from the chains of the intellect and welcoming the sweetness of the unknown is the exhilarating challenge before us. For in this resplendent dawn, every moment is tinged with dazzling light, revealing the unexpected depths hidden beneath fleeting appearances. The blurred contours of existence come to life before our amazed eyes, and the enigma of mystery becomes an inexhaustible source of inspiration and wonder.

Stepping away from the well-trodden paths of rational knowledge, we must lift our eyes to this dawn that illuminates our desperate quest for meaning. Rediscovering the simplicity of mystery means transcending the narrow limits of our understanding and tasting the intoxicating flavour of the ephemeral and the ineffable. Every divine smile offers a glimpse of an elusive truth, and every burst of laughter rings out like a hymn to pure, unadorned joy. As we navigate the twists and turns of life, the serenity of a new dawn beckons us to contemplate the inexpressible with tenderness and humility. It unveils the eternal simplicity of the mystery that embraces us with benevolence.

5

Through the Ages

The Human Quest for Meaning and Joy

Between Ancient Myths and Modern Hopes

An exploration of the invisible threads connecting ancient myths to modern aspirations. Like philosophers in search of truth, we delve into the turbulent realm of human interpretation, navigating the vast expanse between ancient tales and the murky waters of our troubled times. Myths, those ancient tales steeped in mystery and wisdom, act as maps, guiding us towards an understanding of our human condition. Their echoes still resonate today, awakening a deep curiosity in us about the myriad facets of existence. But how can we reconcile these ancient legends with our modern hopes, which are so closely intertwined with doubt and disillusionment? This enigma guides our steps like Ariadne's thread, unravelling through the twists and turns of history. In this endless quest, we search for ultimate meaning, for that fleeting light whose brilliance is reflected through the ages. While myths have nourished the collective imagination for centuries, our modern hopes often seem trapped in the shadows of absurdity. The tension between the grandeur of heroic tales of yesteryear and the cruelty of the contemporary void is palpable, like the discordant notes of an unfinished symphony. Thus, under the piercing gaze of history, we embark on an inner journey to untangle the threads of destiny and free will that are intertwined, searching for the essence of our humanity. To grasp the deeper meaning of our own aspirations, we must explore the limbo of antiquity, embracing the timeless legacy of myths while forging our own path towards the uncertain horizon of tomorrow.

The Desert of the Absurd: Searching for Meaning in the Void

At the heart of human existence lies a barren desert: a sterile territory where any quest for meaning seems destined to fail. It is in this desert of the absurd that man finds himself confronted with the immensity of the void and the emptiness of his existence. Faced with this void, the nagging question of the meaning of life inevitably arises, like a haunting shadow stretching across the endless sands. In this land without landmarks or tangible boundaries, humans struggle, desperately seeking an oasis of clarity and harmony amidst this existential chaos. This desert of the absurd is paradoxical and confusing, defying all logic and reason and plunging the individual into unspeakable fear and profound despair. It is here, amid the ever-shifting dunes of the incomprehensible, that man confronts his own limitations and finitude. Unable to grasp infinity, he wanders lost in this maze of nonsense, desperately searching for an elusive truth. Yet within this relentless immensity, the human soul can discover the seed of an unexpected revelation and a hidden door to freedom. For it is in this very absurdity that man forges his own essence, transcends his condition and, perhaps paradoxically, finds the purest and most authentic meaning. Thus, the quest for meaning in this desert of the absurd proves to be a dizzying adventure: an intimate exploration of the human spirit and a tumultuous journey towards the truth hidden beneath the opaque veils of the universe. Each step in this hostile desert is a step towards a new form of lucidity and a significant breakthrough in understanding oneself and the world. It is through this ardent wandering that a glimmer of hope gradually emerges: a fragile yet resolute promise of meaning at the heart of this desolate immensity.

The Mystical Aura of the Past: Legends and Intertwined Truths

In the twilight of ages past, humanity drank from the fountains of legends and mysteries. Ancestral tales were woven into the fabric of time, lifting a spellbinding mist where echoes of immutable truths mingled with the whispers of elusive myths.

This mystical aura of the past, distorted by time and imagination, still speaks to us today. At the intersection of the tangible and the ethereal, a universe emerges where the line between reality and imagination dissolves in an intoxicating dance.

Legends that have survived the centuries, carried by the voices of storytellers and the pens of poets, populate our collective consciousness. Nestling in the depths of our deepest aspirations, they reveal timeless truths and eternal questions.

Beyond the boundaries of history, legends provide

foundational narratives that illuminate the human quest for meaning and joy. They offer reflections of the human soul, capturing the aspirations, fears, and hopes that transcend time. As we contemplate these ancient tales, we delve into the heart of immutable archetypes, where the contours of eternal humanity take shape.

However, the mystical aura of the past does not merely lull us into reverie. It also lifts the veil on the dark areas of our history, revealing the scars left by unresolved conflicts, abandoned ideals and lost battles. These shadows, sometimes fleeting and sometimes insistent, seal the fate of future generations, compelling them to confront bitter lessons and broken promises.

Thus, in this atmosphere tinged with mystery and timelessness, a complex web emerges, intertwining veiled truths and distorted realities. Between collective memory and individual reveries, the mystical aura of the past continues to fascinate, prompting humanity to explore the depths of its heritage and seek the thread that connects myth and reality.

The Shadows of History: Unfinished Revolutions and Lost Ideals

The tumultuous shadows of unfinished revolts and lost ideals lie in the twists and turns of the past. History is marked by incessant struggles, broken aspirations, and aborted dreams. Each era has had its own torments, truncated revolutions and failed utopias. Noble ideals have been overshadowed by human cruelty and the fragility of social structures. The pages of history are stained with blood and tears, bearing witness to the ongoing struggle for justice and equality. Revolutions have swept across the human landscape, promising freedom and autonomy, but have often been thwarted by the relentless realities of human nature. Martyrs and heroes have sacrificed their lives for ideals that seemed out of reach. The grandeur of popular uprisings only served to highlight the bitterness of resounding failure. Nevertheless, these moments of revolutionary fervour have left an indelible mark on the collective consciousness, inspiring new generations to continue the quest for justice and dignity. Thus, the shadows of history still resonate today, reminding humanity of its past struggles and future potential. In this troubling darkness, however, there is also a glimmer of hope, because every unfinished revolt carries within it the promise of resurgence and every lost ideal contains the seed of rebirth. It is in this twilight that humanity finds the strength to defy adversity, transcend the vicissitudes of time and create a future imbued with justice and freedom.

The Breath of the Renaissance: When Man Rediscovers Himself

The darkness that had enveloped humanity for centuries began to give way under the energetic breath of the Renaissance. Having been plunged into the darkness of ignorance and superstition for so long, man rediscovered himself with a fascinating fervour. The beacons of knowledge, once extinguished, were lit once again, revealing the lost treasures of classical learning.

It was in this dazzling resurgence that a frantic quest for meaning and truth was born. Minds tormented by the rigid dogmas of the old regime embarked on a daring journey of introspection, aspiring to a deeper understanding of their own humanity. The boundaries of art, science, and philosophy dissolved in a whirlwind of tumult, providing a fertile canvas on which humanity strove to decipher the mystery of existence.

Eager to explore the unexplored corners of the human mind, thinkers and artists embarked on perilous expeditions into the heart of the soul. These inner quests led to the rediscovery of the self and the world, giving rise to an unprecedented creative ferment. Through their creations, they sought to capture the vital energy that animated them, shaping a new vision of the world imbued with beauty and truth.

The Renaissance was, in essence, an awakening of the human spirit — a golden dawn after centuries of darkness. It inspired people to take a fresh look at the world around them and question the very foundations of their existence. This fervent search for meaning and fulfilment led to a remarkable flowering of thought, giving rise to innovative ideas that would shape history for generations to come. Thus, the Renaissance was much more than a simple historical transition: it was the defining moment when humanity, like a phoenix rising from the ashes, embarked on its eternal quest to discover its essence.

The Nuances of Reality: Fleeting Joy or Authentic Happiness?

The question of the reality of joy and happiness has haunted humanity throughout the ages. During the Renaissance, man rediscovered his potential and his capacity to experience genuine emotions, including joy. But is this joy merely fleeting and ephemeral, or can it be considered a lasting state of authentic happiness? The quest for meaning in the fluctuating reality of human life confronts us with these delicate nuances. Joy's sparkle often seems to fade in the face of life's torments, dimmed by the shadows of perceived absurdity in every corner of our destiny. Yet there are moments when joy shines brightly, seemingly piercing the darkness of existence and offering a brief but precious reprieve. Is this authentic happiness, or merely an illusory glimmer in the deep night of our lives? Existentialism and absurdism are philosophical movements that explore these questions with striking depth, probing the limits of reality, perception, and human experience. The confrontation between the ephemeral nature of joy and the pursuit of authentic happiness raises fundamental questions about the human condition itself. In this perilous dance between ever-changing reality and timeless aspirations, humans find themselves ensnared by their desires, desperately seeking to grasp the elusive nature of happiness. This perpetual struggle shapes our understanding of reality and our relationship with life itself, pushing us to explore the depths of our being in search of lasting meaning. Thus, the nuances of reality — the difference between fleeting joy and authentic happiness — continue to constitute a fundamental mystery of the human experience, plunging humanity into a dizzying quest for answers that seem to lie beyond the horizon of our most elusive dreams.

The Intimate Connection Between Destiny and Free Will: A Perilous Dance

Since time immemorial, destiny and free will have engaged in an enigmatic dance, weaving the threads of the human story. What invisible force guides our steps, and what role does human freedom play in this grand ball with blurred contours? This question is intrinsic to the very essence of our existence. Sometimes glorified as a transcendent force that governs our lives and sometimes decried as a comfortable illusion that conceals the disturbing reality of our insignificance, destiny has traversed the ages in many forms, defying human questioning. With its inescapable power, it seems to impose itself on individual life paths, providing the perfect excuse for passivity in the face of life's uncertainties. And yet, the rebellious spark of free will emerges, refusing to bow to predetermined paths. Like an echo of freedom in a predetermined symphony, free will charts unexpected paths and unforeseen detours in the course of destiny. But how can these two opposing forces be reconciled? How can chance be granted absolute power while humans are still held responsible for their choices? The confrontation between the immeasurable power of destiny and the indomitable audacity of free will creates a palpable tension that is both an asset and an obstacle to the human adventure. Thus, our quest for meaning and joy is caught between this enigmatic duality. It is in the fragile balance of this perilous dance that the essence of our existence is revealed, oscillating between fatality and responsibility. Contemplating the uncertain horizon of our actions and their repercussions then becomes a profound meditation and an existential quest engraved in the fabric of our being. There is no authentic happiness without this acute awareness of the potential and limitations of our quest; without confronting the dizzying complexity of our condition. Advancing along the winding path of destiny means embracing both the relentless darkness and the dazzling light of our free will; exposing ourselves to both the unalterable thread of fate and the brilliant clarity of our choices. In this perilous dance, we discover the possibility of sculpting our own narrative, of transcending the dictates of fate to compose a symphony that resonates with the breath of our freedom.

Echoes of Cultural Revolutions: Questioning to Move Forward

At the heart of historical turmoil, cultural revolutions sweep in like crashing waves, shaking the very foundations of society. They represent the cry of individuals seeking freedom and defying age-old traditions and established norms. Each revolutionary period is marked by a wind of change, a refreshing breeze that shakes up entrenched certainties and breathes new life into the social fabric.

The audacity of thinkers and artists who bear innovative ideas shakes the foundations of morality and conventional thinking. Cultural revolutions result in creative manifestos and works of art that proclaim the end of obsolete dogmas and the beginning of an era of constant questioning. Art, literature, and music become standard-bearers of an intellectual rebellion, challenging the shackles imposed by tradition.

However, cultural revolutions do more than merely provoke; they raise profound questions about the very essence of humanity and its evolution. They encourage us to look beyond the limits imposed by society and question our beliefs and established ideas. Breaking with the past becomes the catalyst for intense reflection and intellectual renewal, where the clash of ideas leads to collective transformation.

However, this quest for renewal also brings upheaval and conflict, as cultural revolutions cannot progress without friction or resistance. The clash between conservatism and progressivism stirs minds and hearts, creating cracks in the social fabric and fierce struggles over the adoption of innovative ideas. Thus, the path to lasting change is fraught with pitfalls, tensions, and misunderstandings. Nevertheless, it is essential for shaping a society in constant flux.

Despite the turmoil they cause, cultural revolutions embody humanity's vital energy and intrinsic desire to reinvent itself and transcend narrow conventions to open up new horizons. They encourage us to question ourselves constantly, thereby stimulating creativity and innovation. It is through confronting our own certainties that civilisation's ever-changing face emerges: indomitable and constantly evolving, eager to redefine its deepest aspirations in a world in flux.

Towards a New Balance: Ancestral Wisdom and Harmonised Modernity

The upheavals of cultural evolution encourage us to reflect deeply on our path as human beings. Stepping away from the well-trodden paths of tradition and modernity, a new quest is emerging: a burning desire to balance ancient teachings with the necessities of the present. Ancestral wisdom, passed down from generation to generation, now coexists with the frenzy of modernity, magnifying balance and human fulfilment. Achieving this harmony requires uncompromising introspection and respectful dialogue between the past and the future.

Amidst constant societal upheaval, it is crucial that we delve into the depths of our cultural heritage and draw upon the essence of ancestral wisdom. It is in this heritage that the key to inner and outer balance lies. The values passed down by our ancestors — marked by simplicity, respect, and a connection to nature — are beacons lighting the way to a new balance. However, this reconnection with ancient wisdom should not be a nostalgic retreat; rather, it should be an ongoing dialogue between the past and the future.

At the same time, it is crucial to embrace modernity with clarity and discernment. While technology and scientific advances offer new perspectives, they can also threaten to distance us from our deepest essence. It is at this confluence of tradition and modernity that hope for a new era emerges: an era in which humans can fully reclaim their place in the universe. Learning to integrate these two pillars of our cultural identity is the foundation of a balanced and fulfilling life.

Thus, achieving a new balance invariably requires us to question our certainties and prejudices, and to overcome the antagonisms that fragment our perception of the world. Whether in art, philosophy, or spirituality, the encounter between the past and the present reveals new perspectives and unexpected horizons. This relentless quest is the true melody of the human journey and invites us to embrace the richness and diversity of our history, while also projecting ourselves resolutely towards a future built on the foundations of solidarity and kindness.

Conclusion: The Eternal Melody of the Human Quest

We stand at the beginning of a path where the footsteps of past generations still echo and the echoes of their struggles and questions linger. The human quest for meaning and joy emerges as an eternal melody, tirelessly repeated by individuals and societies throughout the ages. This powerful melody's song intertwines with the unique experiences, hopes, and disappointments of the beings who wander the stage of existence. We stand, entranced and perplexed, as this incomplete symphony transcends time and space. For it is in this relentless quest that the essence of our humanity is revealed: an essence shaped by the embrace of ancestral wisdom and the pulse of modernity. Yet perhaps at the heart of this unfathomable depth, we will find the key to piercing the absurdity that shrouds our fleeting lives. Woven with symbols and metaphors, the melody urges us to continue our quest, even when dreams shatter against the cliffs of uncertainty. It whispers haunting harmonies and heart-rending dissonances while offering hope of transcendent revelation beyond earthly torments. Thus, we must sing the symphony of our lives, intertwining our hopes and desires with the threads of destiny and freedom. For it is in this tumultuous dance that our fulfilment and our understanding of the world around us are played out. Like a voyage on the high seas, the human quest punctuates our lives with unexpected treasures, inviting us to drop anchor in harbours of knowledge and fulfilment. Far from being an end in itself, the quest is a perpetual invitation to infinite exploration — an indomitable adventure with no bounds or limits. So, my dear readers, may this conclusion not mark the end of your quest, but rather the start of a new epic. May you become troubadours in search of meaning and joy; frenzied dancers on the stage of the cosmos; and vigilant guardians of humanity's eternal melody.

6

Scientific progress

A God Who Syands Out?

The Long Shadow of Positivism

The influence of positivism on the modern view of science is a question deeply rooted in the foundations of contemporary thought. This philosophy is based on the idea that the only genuine knowledge comes from empirical experience and has had a significant impact on the development of scientific thought. Indeed, it has shaped a vision of science that is focused on rationality, objectivity and the search for tangible, measurable truths. It has established an intellectual framework in which the scientific method is the only legitimate means of acquiring knowledge. Consequently, positivism has instilled a mistrust of anything that cannot be observed, tested, or proven tangibly. While this approach undoubtedly freed science from the chains of superstition and dogmatism, it also created a certain obscurity. By focusing on the visible and demonstrable, positivism has often pushed a multitude of phenomena and perspectives that are essential to a holistic understanding of reality into the background. This obsession with measurable materiality tends to reduce the diversity and richness of human experience, thus limiting the scope of scientific knowledge. In short, the long shadow of positivism has shaped the modern view of science by establishing a clear distinction between the quantifiable and the intangible, and between the verifiable and the speculative. However, this binary dichotomy cannot account for the intrinsic complexity of the universe and the human condition. Therefore, while positivism has undoubtedly brought major advances to science, we must also consider its limitations and consequences in order to achieve a more inclusive perception of reality.

The Light and the Night of Innovation

The contemporary world has become a vast theatre, with the stage dominated by the actors of scientific progress. Innovation seduces with its graceful and deceptive dance, guiding minds towards unexplored horizons. However, it also reveals its dark side, like a starless night where the threatening shadows of technological omnipotence lurk.

Prodigious advances illuminate this tortuous path, giving birth to revolutionary inventions that shape our daily lives. The dazzling glow of these discoveries illuminates our minds with enticing promises, inspiring wonder and a desire to conquer unknown territories. However, lurking behind this dazzling light is an unfathomable darkness dotted with unsettling questions.

Innovation captivates us by offering increasingly daring answers to the mysteries of the universe, but it also plunges us into a sea of questions. It plays with our perceptions, offering radiant visions and dark depths in a fascinating, anarchic ballet.

Confronted with this enigmatic duality, humanity is faced with a dizzying dialectic: the splendour of the human intellect versus the darkness of its unpredictable consequences. Consequently, innovation acts as an elusive muse, inspiring creative geniuses while casting threatening shadows over their work and endangering the fragile balance of our existence. Thus, the uncertain contours of this treacherous path take shape, where the insatiable quest for progress is accompanied by omnipresent dangers, inviting tormented souls to reflect on the pitfalls of their ambitions.

As dawn illuminates the horizon of the future, the flickering lights of innovation offer a contrasting reflection: the promise of a dazzling rebirth mingles with the threat of an inevitable twilight. Like an unfinished epic, humanity continues its journey through the inextricable maze of this relentless quest. We are aware of the dangers that punctuate each step forward, yet we are eager to explore the unknown in the hope of drawing invaluable wisdom from it.

Reflections on the Break with Myth

Does scientific innovation not represent a real break with the myths and beliefs that once shaped our view of the world? The relentless pursuit of truth and rapid technological progress seems to breathe new life into our perceptions, forging connections that are often incompatible with ancient narratives. Tearing away the veil of age-old illusions, these advances shake the very foundations of humanity, confronting it with its atavistic reflexes. Ancestral legends are under attack from the impetuous emergence of objective knowledge, which breaks the age-old veil between the sacred and the profane. This inevitable confrontation between myth and scientific reality therefore raises fundamental questions about human nature and calls into question an entire cosmogony forged over the ages. Ultimately, is man not by nature the bearer of myths and legends, and therefore a mythical being? This unprecedented rupture between ancient fantasy and rationalised knowledge calls into question our very essence, propelling us beyond our comprehension to discover new sources of experience. But should we really disregard our entire mythological heritage in the name of scientific supremacy? The intensification of this tumultuous dichotomy plunges us into the abyss of existential uncertainty, marking an inevitable point of friction between tradition and modernity. A complex picture thus emerges of a humanity whose very foundations are being shaken by the irrepressible impulse of discovery. Could breaking with myth offer the promise of a new horizon where humanity is finally ready to take control of its destiny?

Technological Progress: A New Mythology?

The rapid rise of modern technology has created a new pantheon in which scientific advances become contemporary myths. Technological marvels such as artificial intelligence and biotechnology are shaping a new worldview and shaking the very foundations of our understanding of existence. Like Prometheus offering fire to humanity, modern humans revel in the fruits of their technological quest to elevate themselves to godlike status. However, lurking behind this frenetic pursuit is a disturbing shadow: technology, like the Lernaean Hydra, nurtures both our wildest hopes and our deepest fears.

Technology has become the sacred narrative of our time, revealing a new pantheon of silicon gods whose powers appear to be as immense as their mysteries. Artificial intelligence, like Proserpine held captive in digital limbo, offers the allure of absolute power but also conceals the treacherous trap of insidious dependence. Biotechnology, like the sirens of modern times, promises to rebirth humanity, yet it also stirs deep fears about disrupting the balance of life itself. Faced with these colossal advances, we stand at a crossroads, oscillating between quasi-divine adulation and fierce resistance, like Icarus defying the laws of gravity.

However, amid this technological frenzy, a crucial question emerges: is this new mythology, guided by the quest for progress, leading humanity towards a fascinating peak or the icy abyss of hubris? As silicon gods extend their influence over our lives, we must question the extent of our faith and the omnipotence of the objective mind driving these innovations. Are we truly the masters of this new divine order, or have we created a servitude more insidious than that which we sought to escape? In this technological maelstrom, we must probe the essence of this new mythology to understand its key issues and navigate a path towards enlightenment, free from false chimeras and destructive illusions.

Power or servitude of the objective mind?

The advent of technology today has enabled us to make rapid advances in our understanding and manipulation of the world around us. This quest for objective knowledge has undoubtedly given humanity unprecedented power, granting capabilities once deemed unimaginable. However, as the great thinker Heidegger rightly pointed out, this technological omnipotence is not without risks. When elevated to an altar, the ideal of objectivity could turn into a relentless tyranny, alienating the very essence of humanity. When devoid of ethical discernment, the power of the objective mind runs the risk of turning into servitude, insidiously reducing humans to puppets in the hands of cold, ruthless reason. Thus, we are witnessing dramatic irony: while humans aspire to free themselves from the chains of ignorance, they may well succumb to the shackles of their own creation. This raises the crucial question of human responsibility in the face of this acquired power, born of boundless curiosity and unparalleled tenacity. How can we harness this power for the benefit of humanity, while preserving the integrity of human beings and enabling them to fulfil their spiritual potential? Faced with this challenge, we must give humanity control of its destiny and combine this power with ethical values so that, as masters of our creations, we never become their slaves. It is in this dialectic between domination and freedom that the search for balance is realised, where the power of the objective mind blends harmoniously with an enlightened conscience. This builds a future in which the creature and the creator engage in complete complementarity, and in which humanity asserts itself in all its humanity.

The Insurmountable Limits of Knowledge

Like an unfathomable shadow, the limits of knowledge stand before humanity, challenging our rational minds and our perpetual attempts to penetrate the mysteries of the universe. Beyond the frontiers of the known lies an enigmatic territory where reason wanders without a compass and every truth encountered is accompanied by myriad new questions. The essence of knowledge seems to lie in this infinite exploration, where each discovery brings with it the dizzying burden of the unknown. Scholars courageously venture into these uncertain lands, their minds sharpened by a passion to unravel the mysteries of reality. Yet, despite their prodigious advances, they always stumble upon the elusive boundaries of understanding. Despite the rapid progress of science, the fundamentally limited nature of the human intellect is revealed, hinting at the inevitable failure of any claim to grasp the totality of truth. Thus, like a modern Sisyphus, humanity tirelessly attempts to push the boulder of knowledge to the summit of the mountain of understanding, only to find that the abyss of the incomprehensible stretches mercilessly beyond even the most zealous efforts. The grandiose spectacle of the universe unfolds before our eyes: fascinating and captivating, yet also imbued with a profound ambiguity that constantly challenges reason. In this relentless quest, humans learn that every answer brings new questions and that every certainty tends towards uncertainty. Thus, faced with the abysmal chasm of infinity, human beings are condemned to wander like a modern Prometheus, seeking to unravel the secrets of the gods, knowing their divinity will forever remain out of reach. These insurmountable limits of knowledge reveal the intrinsic modesty of humanity in the face of the immensity of the universe and the tragic grandeur of our eternal quest to understand our own destiny.

Man, architect of his own labyrinth

The path to knowledge is winding and complex, and sometimes paradoxical, yet man persists in his quest for truth. Like a skilled craftsman, he uses reason and logic to create an intellectual labyrinth where each twist and turn reveals new puzzles to solve. This vast web, woven by the human intellect, testifies to its tireless desire to unravel the universe's mysteries. But how can we avoid getting lost in this infinite maze where certainties mingle with doubts? Every step taken in this labyrinth raises a multitude of questions that challenge the very foundations of reason. Man is thus confronted with his own limitations and the fragility of his understanding of the world.

Yet it is precisely in this dizzying journey that humanity's greatness lies. He defies the unknown, delving ever deeper into the realms of thought and scientific discovery. The labyrinth becomes the playground of his mind, where he explores the boundaries of his own existence. At the heart of the paradoxes of existence, he finds his essence — that which distinguishes him as an exceptional, conscious being.

In this labyrinth, he also discovers the relativity of knowledge and its influence on perception. Each new discovery or theory calls into question previously established foundations, casting doubt on acquired certainties. This constant upheaval of human thought solves one enigma only to create a thousand more, thus fuelling the sacred fire of curiosity that consumes the human soul. In the midst of this infinite exploration, humans must accept uncertainty as their companion and not fear the darkness of unexplored passages.

Thus, in his relentless quest for meaning and truth, he becomes the architect of his own destiny. The labyrinth, born of his ingenuity and perseverance, becomes a symbol of his struggle to comprehend the incomprehensible. It is precisely in this ceaseless creative act that the essence of his humanity lies. Although the labyrinth may seem impenetrable, man continues to face mysteries with fearless determination by his very nature.

Paradoxes of a Scientific Odyssey

In our relentless quest for understanding and progress,

the scientific odyssey unveils intriguing paradoxes, enveloping humanity in a maze of contradictions. On the one hand, each advance illuminates the dark corners of the universe, opening up infinite horizons of discovery and innovation. But, on the other hand, these flashes of light also cast disturbing shadows, revealing unfathomable abysses and elusive enigmas. In this kaleidoscope of ever-evolving knowledge, the foundations of truth seem to blur. As we scan the stars in search of our origins, the deeper we delve into cosmic mysteries, the more elusive our place and role in this vast universe seem to become. The formidable power of human reason appears to collide with the insurmountable limits of knowledge. Thus, each step forward reveals the inextricable knots that weave the fabric of our existence. The result is a paradoxical portrait of modern man: a surveyor of infinite knowledge, yet confronted with his irremediable ignorance. This constant dance between the dazzling light of discovery and the enigmatic darkness of the unknown creates an existential tension that shapes our relationship with the world. How can we invoke rationality without forgetting humility in the face of the immense mysteries that remain? How can we embrace the momentum of progress without being blinded by our own creations and drifting towards the decay of the human essence? These questions reflect our condition and lay the groundwork for a metaphysical exploration at the heart of the scientific adventure. Thus, the paradoxes of an odyssey in which humanity seeks to leave its mark on the universe while being marked by its discoveries are revealed.

Between Radiant Promise and Decay

The rise of scientific progress, like an elusive prodigy, weaves a dazzling tapestry of infinite promise. The dazzling light of knowledge ignites hearts and awakens in humans an insatiable thirst for understanding and unravelling the immemorial mysteries that shroud our existence. However, behind this radiant facade, the menacing shadow of silent decay is quietly taking shape: a pernicious deviation in the tumultuous course of progress.

While the prospect of interstellar travel, miraculous cures and instant communication seems to promise a bright future, there is a sinister side to this. A form of progressive dehumanisation is taking root in our societies. Humankind is becoming enslaved by its own creations, bound by technology that threatens to confiscate free will and transform life into a series of fleeting moments lost in the illusion of permanent connectivity.

The beauty of scientific progress, like a mirage promising an idyllic future, is confronted with the reality of its decline. The vertigo of omnipotence it inspires is accompanied by the latent danger of self-destruction, which is inherent in human arrogance. Are we creating a new mythology where our irrational faith in reason clouds our judgement, exposing us to the risk of a dizzying ethical decline?

Our insatiable quest for progress is tinged with a paradoxical duality of fulfilment and peril, challenging our humanity on an uncertain path. On the horizon of this scientific odyssey, the radiant face of discovery meets the growing darkness of carefully concealed decay. Therefore, let us scrutinise this symphonic arrangement of sparkling hopes and insidious threats with apprehension. Its fertile dynamics will shape the ethics of the future, necessitating an enlightened responsibility to pass on tomorrow's heritage intact.

The Awakening to an Ethics of the Future

In the shadow of scientific progress, the crucial emergence of constantly evolving ethics is taking shape. The future horizon, once defined by ancient narratives and beliefs, is now animated by the unexpected potential of science. At the dawn of this new world, awakening to an ethics of the future necessitates profound introspection and a compassionate, reflective approach to the future of humanity.

At the heart of this reflection resonates the powerful echo of existential questions. Therefore, scientific progress cannot be understood outside of an ethical context. It demands the responsibility of each individual to help shape our collective future. Thus, the complex fabric of an emerging ethics is taking shape, striking a balance between seductive promises and latent dangers.

From this perspective, there is an urgent need for collective awareness. The ethics of the future are emerging through open dialogue between science, the arts, and the humanities. This dialogue is nourished by fruitful debates on the socio-economic, political, and ethical implications of technological advances. In this sense, the seeds of future ethics transcend disciplinary boundaries, encompassing the full range of contemporary challenges and offering insight into the rapid changes around us.

Nevertheless, the quest for an ethics free of illusions requires unwavering vigilance. While scientific progress promises to emancipate humanity, it also carries within it the seeds of desolation. Against this backdrop of delicate balance, multiple voices converge on an urgent need: to forge a robust and enduring ethics capable of guiding humanity's hesitant steps towards a horizon that offers true reassurance.

It is in this spirit that the urgent need for international cooperation emerges. The ethics of the future extend beyond geographical and ideological boundaries, forging new connections between nations and peoples. They call on each of us to consider our place and responsibility within this vast human tapestry.

At this pivotal juncture, our shared destiny is revealed amidst the glimmers of emerging knowledge, the palpable uncertainty of impending choices, and the reassurance of ethics as our ultimate refuge.

7

Painting at the Heart of Man

Autonomy of the Soul through Creation

Man and Canvas: The Ink of the Inner World

In the dark studio, where shadows dance to the rhythm of brushstrokes, the artist stands facing the blank canvas. It is in this silent intimacy that the secrets and torments of the human soul are revealed. Perched on his easel, the artist's discerning gaze captures the shifting colours of the soul, capturing every nuance as emotions unfold. Like an alchemist, he translates these invisible emotions into a striking chromatic symphony, revealing inner landscapes that are often hidden or impossible to express by other means.

Each brushstroke is a quest for transcendent truth, an attempt to capture the essence of existence itself. Bright flashes and dark hues intertwine on the canvas like silent confessions engraved in the eternity of art. Like an intrepid explorer, the artist delves into the depths of the human soul, unveiling previously unknown inner landscapes.

Through this communion with the canvas, humanity finds a way to transcend its condition, rising above material contingencies to give shape to its dreams, fears, and hopes. Art becomes a mirror of the soul, reflecting inner conflicts and personal triumphs and offering a true window into the human psyche.

Each piece is an invitation to immerse oneself in a parallel universe — a realm where words are superfluous and only emotions have a voice. Like a shaman performing a mystical dance before the altar of creativity, the artist transmutes the torments of the soul into a visual symphony, offering salutary catharsis for himself and those who contemplate his work.

Thus, the canvas becomes a sanctuary where man can exorcise his demons, express his joys and reveal the depths of his being. It enables him to give form to the ineffable and defy the unspeakable.

Colours of the Soul: Palette of Invisible Emotions

The infinite palette of human emotions is poured onto the canvas — a volatile and enigmatic mixture that transcends words. Colours, the hues of the soul, express the richness and complexity of the human experience. Each shade reveals an inner universe, a subtle reflection of the torments and ecstasies that inhabit the human soul. Under the artist's brush, these colours come to life, intermingling to create paintings that vibrate with emotion. The fiery red of passion, the deep blue of melancholy and the dazzling yellow of joy each evoke a unique set of indescribable feelings and impressions. Just as a musician tunes their instrument to compose a spellbinding symphony, the artist wields their colours with finesse to translate the ineffable. The palette of invisible emotions offers an escape from rational thought, plunging the mind into an ocean of intuition and elusive sensation. It invites observers to transcend tangible reality and discover the depths of the human soul. These evocative nuances weave a universal emotional language that transcends linguistic barriers. They forge an intimate bond between creator, work, and viewer, creating a complex web of mutual understanding. Far from being mere pigments applied to canvas, the colours of the soul are messengers and mediums through which artists communicate their deepest essence. They embody the essence of humanity, revealing an infinite spectrum of feelings that are sometimes unfathomable. Thus, emotions become tangible entities, transcended by the magic of art to touch hearts. Each brushstroke is an invitation to embark on an introspective journey and explore this intimate universe where unspeakable suffering and ineffable ecstasy coexist. Colours vibrate in unison with the heartbeat, capturing the transient expressions of the soul in a mesmerising dance on the canvas. They remind us that behind every shape and line is the echo of immortal emotions that shape our humanity. Like an alchemist of emotions, the artist magnifies the palette of his soul to illuminate those of others, offering a secret passage to the raw truth of existence.

Behind the Mask: The Autobiographical Portrait

The saga of the human soul is drawn in every brushstroke and line. Each work is a mirror in which the artist reflects himself. This autobiographical portrait is an exploration of the deepest recesses of the mind, far from being a simple physical representation. The artist, both witness and participant in their own drama, unveils the secrets buried deep within their being. Every nuance, contour and reflection offers a dizzying plunge into the artist's tormented psyche. Then, the work comes to life, taking on the appearance of introspection, which is both dream-like and feared. Inner demons dance across the canvas amidst interplay of light and shadow, creating a bold fresco in which the self confronts its contradictions.

Through this canvas, the artist rebels against the determinism that binds individuals to predetermined destinies. They break free from the boundaries imposed by society, seeking to breathe life and freedom into their identity. Each brushstroke is a fierce struggle against the inevitable, a silent cry that screams the will to create oneself beyond social constraints and normative pressures. Thus, the pictorial autobiography becomes a battlefield where the artist combines strength and vulnerability to forge his own reality, free from the shackles of a predetermined destiny.

Behind the mask of colour and form lie the most intimate layers of the soul, exposing the raw truth in all its splendour and nakedness. Each brushstroke is a confession, an admission and an affirmation of the self, capturing the ever-changing nature of human identity. In this emotionally charged self-portrait, the quest for self intertwines with the desperate search for inner coherence, reminding us that art is a visual meditation on the enigma of human existence. In this incessant ballet, which the artist calls creation, he seeks to grasp the tenuous thread that weaves destiny into a worthy fabric.

Creation versus Destiny: The Struggle for Identity

In our relentless quest for our essence, we face an eternal dilemma: the choice between the crushing weight of destiny and the transformative power of creation. The very notion of destiny seems to freeze our existence in a preconceived pattern, placing us in predetermined roles from which there is no deviation. However, artistic creation offers an escape, a path towards reaffirming our deepest identity. By defying the constraints of destiny, artists free themselves from the grip of predetermination, asserting their autonomy in an act of pure freedom. Each brushstroke and line on the canvas becomes an affront to destiny and an affirmation of the artist's individuality in the face of convention. However, this struggle is not without its challenges. It exposes the artist to uncertainty and confrontation with the doubts and torments that accompany any act of rebellion. The struggle for identity through creation is a risky endeavour, a battle against established norms and unquestioned dogmas. Nevertheless, it is precisely in this struggle that we discover the essence of our humanity: the refusal to submit to a predetermined destiny imposed by external forces. By taking control of their destiny and giving shape to their deepest aspirations through art, individuals transcend their condition, revealing themselves fully to themselves and to the world. This raises the crucial question: can creation truly defeat destiny, or does it merely give us the illusion of a fleeting victory? It is in this perpetual confrontation that the drama of human existence plays out: creation becomes the means by which we resist the annihilation of our identity and find respite in the infinite dance of destiny.

The Shadows of Existence: Pictorial Interpretations of the Void

Human existence, like a blank canvas before a painter, can be interpreted in many fascinating ways. Its infinite complexity casts enigmatic shadows, inspiring pictorial representations imbued with profound meaning. Like veiled reflections of the human condition, these shadows come to life on the artist's palette, revealing truths buried deep within the unconscious.

The underlying quest for meaning is evident in the subtle gradations where dark and light shades intermingle. Courageous artists dare to explore uncharted territories of suffering and loneliness, capturing the ephemeral and elusive with impeccable grace. The void — that abyssal space which haunts tormented souls — becomes a subject of boundless exploration, revealing the most intimate intricacies of the human psyche.

Imagery of shadows transcends the mere interplay of light and darkness, revealing the essence of existence itself. In emotionally charged paintings, each shade and each stroke becomes a living metaphor for our earthly existence. This ambiguous and disconcerting void, ever-present, shakes established certainties, prompting the observer to delve into the heart of these representations to uncover a fragment of elusive truth.

It is in this exploration of the shadows of existence that the true essence of humanity is revealed. The artist becomes an alchemist, shaping the raw material of their emotions to create works of overwhelming intensity. Each painting offers a different perspective, creating a dialogue between the work and the viewer and inviting the latter to explore their own depths with courage and clarity.

Thus, beyond the canvas, an invitation emerges to contemplate the inexpressible and embrace the ambiguity of our ephemeral existence. Points of reference waver, contours blur, and we find ourselves confronted with the unfathomable mystery of our finite existence. In this pictorial meditation, hope emerges: fragile, yet resolute. It invites each of us to cross the threshold of shadow and find fulfilment at the heart of emptiness.

The Revolt of Forms: A Declaration of Freedom

Art, in its purest form, is a revolt against conformity and oppression. Each brushstroke and pencil line affirms freedom and proclaims authentic existence in the face of the sterile banality of everyday life. Through the chaotic interplay of shapes and colours, the artist seeks refuge not to escape the world, but to gain a deeper understanding of it and transcend its limitations. Every painting, sculpture, and abstract composition is a call for emancipation and a challenge to the established order.

When an artist confronts a canvas or any other surface for expression, it is an act of rebellion against the invisible chains attempting to restrict human expression. Through its tormented lines, striking contrasts, and unexpected perspectives, a work of art becomes a bastion of freedom — a territory where the judgement of others has no place. By breaking with convention and freeing themselves from social expectations, artists proclaim their own elusive yet palpable truths through their compositions.

By exploring rebellious forms, the artist challenges the limits of the visible and invisible worlds. Each composition is a declaration of the infinite potential of human creativity and a bold affirmation that imagination cannot be tamed. Swirling shapes, untamed contours, and empty spaces express the perpetual quest for meaning and the endless exploration of the human soul. It is in this rush towards the unknown that the artist finds their true essence, free from the constraints of material reality and rational logic.

The revolt of forms will always be a courageous affront to the darkness of existence. Each artistic composition offers a glimmer of hope and affirms that creation can transcend the triviality of the material world. Through their heroic works, artists invite viewers to join this eternal rebellion, to embrace freedom of expression and to assert their existence beyond the boundaries of everyday reality.

Liberation through Expression: Beyond Perceptible Boundaries

Freedom, an intangible yet compelling concept, bursts onto the canvas like a powerful wave. Like a universal cry, artistic expression dissolves the perceptible boundaries between here and there, the self and the other, the known and the unknown. Each brushstroke and speck of pigment becomes a silent cry that resonates beyond contingencies and the constraints of everyday life. During the creative process, the artist transcends their own limitations and explores the infinite depths of the human mind.

The artwork, a silent witness to this inner journey, opens a window onto the unimaginable, inviting the viewer to immerse themselves in an ocean of pure emotion. Colours intermingle and shapes intertwine, and the artist, like an alchemist, distils their torments and ecstasies into a visual symphony that only ardent hearts can hear.

Beyond the boundaries of the senses, artistic expression unveils the paradoxical reality of universality within singularity. Each stroke bears the imprint of a personal story that resonates with the stories of thousands of other souls. Thus, each creation becomes a gateway to both the intimate and the universal — a fragile yet indestructible bridge between individuals.

In this uninterrupted exploration, the artist moulds space and time at will, creating a new geography where physical laws fade and creative momentum alone prevails. The mind is freed from all constraints, crossing the barriers of tangible reality and rising towards unlimited horizons.

What can be said about the fertile mirage that is a work of art? It is a source of wonder and a mirror that highlights the hidden tears of the human soul. Through the eloquence of its silence, it celebrates the beauty and fragility of our condition, reminding us of the infinity within our finitude.

From the Silences of the World to the Eloquence of Art

Silence is the music of infinite spaces, where thoughts dance with the solitude of the human soul. It is in these silences that art finds its true eloquence, asserting its presence beyond words and discourse. When the world holds its breath, canvases, pieces of marble and musical scores become powerful ambassadors for expressing the inexpressible.

Like alchemists of emotion, artists capture these silences and transform them into works of indescribable beauty. Each brushstroke, musical note and sculpture shapes a language that transcends rational barriers to reach the heart of humanity. These creations are not merely aesthetic artefacts; they are portals to the deepest recesses of the human psyche. They delve into the depths of the human soul, awakening indescribable emotions and evoking a universal resonance.

In this subtle dialogue between the silence of the world and the eloquence of art, a poetry without words emerges: a dance without movement; a symphony without sound. Silences become the solid foundations on which humanity's masterpieces are built, carrying timeless messages. They open doors to unexplored inner worlds, inviting everyone to delve into the depths of their own existence and find an echo in these immortal creations.

The eloquence of art lies in its ability to transcend the limitations of verbal language and speak to the soul. Every line drawn, every shape formed and every harmonious composition is a silent cry that transcends cultural, linguistic and temporal boundaries. In this way, art reveals its unifying power, connecting individuals beyond material contingencies and offering them refuge when words are inadequate and the turmoil of the world subsides.

As spectators of this eloquence, it is then up to us to listen attentively to the language of silence, to let ourselves be carried away by the brilliance of artistic expression and to surrender to the rich meaning of every silence transfigured into a work of art. For it is in these privileged moments that we access the universality of our condition, transcending personal contingencies to join the vast ocean of humanity.

Encounter with the Absolute: At the Convergence of the Imaginary and the Real

In this workshop of colours and shapes, we witness the dazzling union of the Absolute and the tangible, the imaginary and the real. The canvases come to life, revealing the deepest secrets of the human soul and capturing fragments of reality that blend harmoniously with abstract dreams. Each brushstroke is a bold statement, seeking to transcend the boundaries of the known world and explore the unknown with burning passion. The artist confronts the enigma of existence, evoking elusive truths that defy logic and inspire endless contemplation.

At the heart of this fertile convergence, a mystical spectacle unfolds where creative effervescence bursts into an intimate dialogue between mind and matter. Like a fervent prayer, the power of imagination gives birth to visions of transcendent beauty. Every glance at these works of art is a step into a parallel universe, where reality and fantasy dance together in an intoxicating sarabande. By transcending the boundaries of ordinary perception, art becomes a guide to unexplored territories where the soul can find refuge in its eternal quest for meaning.

On this mind-blowing journey into the absolute, the artist becomes an intrepid messenger, exploring the depths of existence and offering viewers the chance to reach heights that would otherwise be inaccessible. Each piece opens a door to a metaphysical dimension and invites viewers to contemplate infinity through the eyes of the creator. These encounters with the absolute evoke a multitude of emotions, ranging from wonder to perplexity, and offer viewers a privileged insight into the profound nature of humanity. Some sense the myriad possibilities revealed as they gaze at these paintings, while others allow themselves to be enveloped by a meditative tranquillity, rekindling their connection with the mystery of the universe.

Thus, the convergence of the imaginary and the real offers a symphony of ideas and perceptions that constantly redraw the fabric of our existence. Within this artwork, humanity finds refuge in its struggle against oblivion, building bridges between the ephemeral and the eternal. Each canvas becomes a gateway to the vastness of possibilities, inviting curious minds to immerse themselves in the fascinating complexity of this timeless encounter.

Painting the Future: Utopian Visions or Mirrors of Reality

The evolution of painting is part of a timeless dialogue between possibility and impossibility, utopia and reality. In each brushstroke lies a projection into the unknown, an exploration of uncharted territories beyond the fixed contours of the present. It is in this tension between dream and truth that the uncertain horizon of the future takes shape — a horizon that the artist tirelessly attempts to anticipate through their canvases.

Utopian visions shine through in landscapes that are

wonderfully impossible, where the sky is adorned with surreal colours and nature rises up in an unexpected symphony of shapes and shades. Each painting becomes a reflection of an alternative universe and an invitation to explore the unexpected corners of the human mind. These creations are far from mere chimeras; they shake the very foundations of perception, taking the viewer on a journey through space and time where the boundaries of the possible are erased.

However, beyond this escape to unexplored worlds, painting can also offer reflections on the reality that surrounds us. Through striking representations of familiar scenes, the artist uncovers the buried truths of our human condition. Each brushstroke becomes a poetic reminder of our joys, struggles, and disillusions, opening doors to a deeper understanding of our existence.

This duality between utopia and reality is embodied in each masterpiece, revealing a part of the human soul steeped in grandiose aspirations and earthly roots. In this infinite exploration of possibilities, painting becomes a beacon guiding humanity towards a future where the boundary between the imaginary and the tangible dissolves, leaving only the dazzling harmony of creation.

8

The Embrace of the Religious Community

A platform for hope or confinement?

The Illusory Promise of Unison

The mirage of unison within a religious community reveals a striking duality. On the one hand, there is the comforting promise of belonging to a coherent whole where differences fade and hearts beat as one. This illusion intoxicates souls in search of comfort and certainty, offering a refuge where the torments of uncertainty dissolve. However, behind this apparent unity, subtle coercion can lurk, taking the form of rituals and symbols that bind the faithful in a deceptive uniformity. The desire to maintain this artificial harmony can stifle diversity of thought, numbing minds under the weight of prescribed conformity. Thus, the much-desired unison becomes an insidious trap, offering not true communion, but a parody tinged with docility. Many oscillate between the lure of feigned unity and the muffled cry of individuality in search of freedom, caught in this precarious balance between comfort and oppression. This deep tension gives rise to existential questions, confronting believers with the difficult duality of finding comfort within a group while preserving their unique essence. Therefore, unity within the religious community is a shifting promise, oscillating between the warmth of illusory cohesion and the chill of disguised constraint.

Rites and Symbols: Invisible Bonds or Chains?

Rites and symbols, born of age-old traditions, appear to forge invisible bonds within religious communities. Steeped in meaning and history, these ritual practices offer a symbolic dimension that transcends everyday life. Celebrations punctuate the lives of the faithful, creating a space where the individual and the collective fuse. However, beneath this enchanting aspect lies a darker reality.

While rituals can initially strengthen a sense of belonging and promote the transmission of sacred values, they can also become invisible chains that hinder freedom of thought. They can freeze minds in rigid attitudes, relegating the individual to the background in favour of the community's prestige. Blind adherence to these rituals can lead to uniformity of thought and the neglect of important questions.

Furthermore, symbols carry intense emotional weight and permeate the collective imagination, helping to forge a common identity. However, behind this attractive facade, rigid dogmas can lurk, acting as hidden chains that weigh down individual thought. Thus, symbolism, far from being purely aesthetic, can imprison the mind within fixed representations, thereby restricting the capacity for reflection and questioning.

Faced with this ambivalence, the question remains: how can we reconcile the enriching contribution of rituals and symbols with the necessary liberation of the individual? How can we break free from these invisible chains while preserving our cultural and spiritual heritage? These questions demand deep reflection and open dialogue about the evolution and relevance of these practices within religious communities.

The Sacred Under the Mask of Community

When we examine religious practices closely, we discover a complex web in which the sacred is hidden beneath the veil of community belonging. Celebrations and rituals seem to weave invisible bonds that anchor individuals within the spiritual community. For some, these customs take on a transcendent dimension, transporting them to metaphysical realms where individuality gives way to a sacred union. However, behind this mask of the sacred, the insidious shadow of social control sometimes lurks. How does sacredness become the dark ferment of uniformity? The answer lies in the very nature of human beings, who hunger for meaning and transcendence. Religious communities offer a language, symbols and myths that colour existence with divine hues, but these can also die, freeze or oppress. For beneath the garb of holiness often lurk figures of power. These mystical authorities embody revealed truth and hold the privilege of interpreting divine mysteries. In this way, they establish a moral and social order, structure the collective imagination, and shape individual consciousness. But how much of this authority is legitimate, and how much constitutes abuse or oppression? This remains a thorny question, as injustices have been committed, minds have been subjugated and souls have been condemned in the name of the sacred. Nevertheless, the desire for communion, exchange and transcendence persists. It is therefore essential to distinguish the authentic sacred from the communal trap, offering everyone the possibility of journeying towards the ineffable with complete freedom and without alienation or domination. Thus, freed from the invisible chains of orthodoxy, human beings may be able to embrace their spiritual quest with renewed clarity and rediscover the sacred beauty that unites us all beyond dogma and constraints.

Figures of Power: Spiritual Authority Unmasked

In the perpetual quest for meaning and unity, spiritual authority is often cloaked in mystery and sacredness. However, behind this sacred façade, figures of power with questionable motives sometimes emerge. These authorities, who are supposed to embody purity and wisdom, sometimes have a much more earthly side, marked by desires for control and domination.

Spiritual authority unmasked thus reveals the power struggles that often underpin the very foundations of faith. Those who wield this authority present themselves as guardians of dogma and tradition, but sometimes conceal their own ambitions behind a veil of transcendence. Their words, once revered as absolute truth, are now tinged with disturbing nuances of human power.

These figures exert a profound influence over the faithful, drawing them into their sphere. Their role as spiritual guides insidiously transforms into an instrument of control, stifling all forms of questioning and free will. Thus, instead of liberating consciences, spiritual authority can lead to their alienation, reinforcing the invisible chains of submission and unconditional devotion.

Yet, despite these dark shadows, history has also seen the emergence of authentic spiritual figures guided by compassion and a sincere quest for truth. These beacons of humanity offer an alternative to authoritarianism, prompting deep reflection on the nature of spiritual authority itself. Their humility and empathy reveal that true power lies in the ability to free souls rather than enslave them.

Therefore, deconstructing spiritual authority involves probing the depths of the human soul, questioning the labyrinths of faith, and tracing the shifting contours of inner freedom. It is in this relentless pursuit that the paradoxes of humanity in the face of the intangible are unveiled, in an endless dance between adoration and resistance, between spiritual fulfilment and subjugation.

Mystical Exaltation: The Dance of Faith

Like a celestial choreography, the dance of faith guides believers towards a state of spiritual ecstasy where the divine merges with the human soul. This mystical exaltation transcends material limits, elevating the spirit to immaterial spheres. When sacred music fills places of worship, it resonates with every fibre of the faithful's being, drawing them into an intoxicating trance that abolishes the boundaries between the earthly and the divine.

In this harmonious dance with the universe, believers experience deep communion with the sacred. Ritualised gestures become a physical manifestation of inner devotion, while chanted songs provide a spiritual soundtrack. Every movement and every note is an offering to the deity — a prayer embodied in physical expression.

However, mystical exaltation raises questions. At the heart of this intoxicating dance lie the paradoxes of faith. The quest for transcendence can clash with institutionalised dogmas and strict rules that attempt to define the indescribable. Individual aspirations for inner freedom may conflict with the norms established by the religious community, generating dissonance between the search for the absolute and the constraints of dogmatism.

This tension between mystical ecstasy and existential dilemmas gives rise to profound questions for believers. How can the freedom of the soul be reconciled with ritual constraints? And how can the purity of the mystical experience be preserved in the face of tradition's rigidity? These questions lie at the heart of the dance of faith and reveal the intimate struggles of souls in search of truth.

Thus, the dance of faith embodies the fundamental conflict between the individual's spiritual quest and the collective framework of religious practices. It is in this dynamic tension that the bonds between believers and their quest for transcendence are formed and transformed. Mystical dance is a powerful expression of human restlessness in the face of the infinite, oscillating between inner freedom and social restraint, and between personal transcendence and belonging to a community.

Modern Dissonances: Conflicts between Dogma and Freedom

In the tumultuous arena of modern society, ancestral dogmas clash with aspirations for individual freedom. The truths established by religious institutions are constantly challenged by an insatiable quest for freedom of thought and action. This bitter conflict between tradition and modernity, between the sacred and the profane, is shaking the very foundations of our societies. On the one hand, the guardians of the faith defend immutable precepts, proclaiming the infallibility of the dogmas that guide them. On the other hand, free spirits demand the right to autonomy of thought, rejecting all forms of intellectual imprisonment. This bitter struggle creates a climate of discord where every divergent thought is perceived as a threat. The legitimacy of dogma is being undermined by rapidly changing attitudes and customs. Every advance in science, technology, and critical thinking undermines age-old certainties. Yet, at the heart of this confrontation, there are voices emerging that seek reconciliation and invite us to rethink the relationship between dogma and freedom. Should we sacrifice the individual on the altar of age-old traditions, or allow them to develop their intellectual and existential capacities fully? This remains a burning question, fuelling heated debates about the place of religion in the modern world. Beyond the divisions, there is an urgent need for constructive dialogue to enable peaceful coexistence between different perceptions of the sacred and the secular. Whether it involves rethinking sacred texts in light of contemporary advances or accepting the harmonious coexistence of multiple truths, there are plenty of solutions, but they require open-mindedness and shared humility. Only deep and collective reflection can lead us towards a new balance between dogma and freedom, enabling us to build a world in which everyone can find their place without hindrance or alienation.

Les Espoirs Enchaînés: Regards croisés sur l'enfermement

At the heart of human consciousness, the fragile threads of hope and disillusionment are intertwined. In this timeless essay, we examine with philosophical rigour and deep compassion the invisible chains that captivate the distraught soul. Such is the fate of humanity when confronted with the concept of the divine as proclaimed by religious institutions. How can we untangle the web of beliefs, hopes, and disillusions that imprison the mind?

Faced with this tangle, we see the sorrowful faces of souls seeking refuge in spiritual communion. Their songs of praise echo in sacred spaces, enveloping kneeling figures in solemnity. But behind this majestic facade lie muffled whispers of stifled hope. The glances exchanged in this subtle confinement provoke inner turmoil and raise profound existential questions.

This heavy silence is broken by the liberating words of apostates and heretics — daring dissidents who defy dogmatic authority. Their impertinent words resonate like a discordant symphony within these spaces, challenging the unshakable order of tradition. Yet, in their iconoclastic tales, we discern a courageous struggle to cast off the mental yoke and embrace the light obscured by the shadow of dogma.

These contrasting perspectives on confinement reveal the tension between freedom and submission. The tormented mind oscillates between an ardent call for freedom and a deep-rooted fear of the unknown. These hopes, chained together, are silent witnesses to an inner, underground struggle where the quest for meaning intertwines with the aspiration to break free from the invisible bonds that hold the soul captive.

Beyond these gazes, the insatiable quest of the soul in search of redemption emerges. The chains of confinement melt under the fiery heat of introspection, revealing an inextricable paradox: hope, both captive and jailer, becomes both the burden and the escape of the tormented soul. This introspective dialogue ends, like a poignant lament, illuminating the winding paths of spiritual imprisonment.

The Escape from Silence: Words of Apostates and Heretics

The silence of beliefs imposes its tyranny on souls in search of freedom. The words of apostates and heretics dance like bright flames in the darkness of religious confession. Their stories, written by those who were excluded, reveal a truth that is often stifled by dogma. These dissident voices, with their quiet accents of rebellion, initiate the process of intellectual emancipation.

Like a sun hidden behind opaque clouds, apostates shake the very foundations of faith. Their testimonies are beams of light, piercing the thick fog of religious tradition. Free from the shackles of outdated orthodoxy, these luminous misfits challenge the authority of the clergy and overturn long-held certainties.

Meanwhile, heretics stand as sentinels at the edge of established dogma. Their words, like dissonant notes in a celestial symphony, reveal the cracks in the dogmatic structure, which has been shaken by time. They demonstrate the scope of their independent thinking, offering unexpected possibilities like a clear horizon.

Through their confessions, these marginal figures offer readers a dizzying descent into the unexplored depths of rebellious thought. Their words, laden with secular echoes, crystallise the emergence of sacrilegious questioning in the face of immutable truths. Each of their diatribes, woven from threads of audacity and irreverence, exudes the intoxicating scent of freedom of thought.

The words of apostates and heretics are odes to difference, challenging the dogmas that create divisions between people. They sketch the contours of a landscape where every mind can roam freely without fear of excommunication or divine retribution. Thus, the voices of those who were once silenced rise up to shape a subversive polyphony tinged with hope, where dissent becomes self-affirmation and divergence a promise of awakening.

Ephemeral Utopias: When Cathedrals Tremble

The world of beliefs is a theatre in which many dramas play out, both individually and collectively. Among these striking representations are sacred buildings, true symbols of power and permanence. Cathedrals, architectural jewels built through faith and human ambition, stand majestically within urban landscapes as silent witnesses to successive generations' hopes and torments. Yet, behind their imposing façades and celestial vaults, there are sometimes fleeting illusions and unexpected fragility. Let us contemplate these stone colossi that falter when certainties waver and the foundations of a society in search of meaning are shaken. Cathedrals, seemingly eternal, are nevertheless vulnerable to the turmoil of history. They bear the scars of spiritual change, internal revolt, and disillusionment — symbols of the impermanence of utopias. Every crack in their architecture echoes ideological conflicts and social divisions, revealing the transient grandeur of these sacred art wonders. When cathedrals tremble, it is an entire worldview that wavers. The illusion of permanence dissipates, revealing a more changeable and uncertain reality. Centuries of dust cover the capitals, yet the thirst for ideals remains unquenched. Despite the tremors and storms, the quest for a timeless refuge — a spiritual haven in which to find rest and inspiration — endures. Thus, ruined cathedrals provide a magnificent backdrop for new struggles, unexpected awakenings, and salvific metamorphoses. Another kind of grandeur emerges: perhaps less spectacular, but more authentic. It reconnects with humanity and transcends stone to exalt the spirit. These buildings, symbols of a bygone era, nevertheless keep the flame of human aspiration alive, breathing new life into the heart of the ruins. When cathedrals tremble, unexpected promises are born — fresh hopes that glorify the ephemeral and celebrate fragility as the very essence of splendour.

Rediscovering Humanity: The Path to a New Awakening

Humanity is the cornerstone of our existence, yet it often finds itself lost in the maze of modern society. But what does it mean to rediscover humanity? Is it a utopian quest or a reality within reach? Perhaps the path to a new awakening lies in rejecting the artifice of our contemporary world and reviving what is essential: what makes us authentic, conscious beings.

In our relentless search for meaning, we must question our lifestyles, values and beliefs to gain a deeper understanding of ourselves. Freeing ourselves from the shackles of conformity and breaking out of social constraints are undoubtedly the first steps towards a new dawn, where individuality and authenticity can flourish.

This renewal requires a return to otherness, embracing the uniqueness of others. Far from standardisation and uniformity, it is by embracing human diversity that we can cultivate this long-awaited awakening. This means breaking down the barriers of indifference, relearning the language of compassion and solidarity, and building a space of brotherhood and mutual respect together.

However, rediscovering our humanity cannot be achieved by ignoring the spiritual aspect of our being. In this quest to reconnect with our humanity, spirituality is paramount. This exploration should not be confined to a pre-established religion, but rather be an intimate journey of discovery into our relationship with the divine and the universal. Moving beyond fixed dogmas and embarking on a free and personal journey seems to be the key to transcending our limitations and rediscovering our deep humanity.

Beyond the turmoil and upheavals of our time, rediscovering our humanity takes on a significant existential dimension. It is a call for reconciliation with ourselves, with others, and with the universe as a whole. This path towards a new awakening offers a glimmer of hope: a path towards collective rebirth, where humanity — in all its vulnerability and greatness — once again becomes the essential pivot of our shared destiny.

9

Inner Revolution

When Science and Art Shape the Mind

The Awakening of Consciousness: Entering the Modern Soul

In a world of constant change, the modern soul undergoes personal awakenings that profoundly alter its perception of existence. Navigating this maze of change, individuals embark on an unprecedented inner quest where consciousness gradually blossoms. While this inner transformation is challenging, it also offers the chance to break free from the dogmatic beliefs inherited from the past. Thus, the modern mind becomes aware of its ability to think beyond established dogmas. This liberation of thought becomes the driving force behind profound personal evolution, where every preconceived idea is rigorously scrutinised. It is through this questioning process that the modern soul finds fertile ground for its awakening and is invited to explore previously unexplored paths. Venturing far from the beaten track, consciousness opens up to new perspectives and awakens to a renewed understanding of its own essence. By defying established norms, the modern individual experiences a radical inner revolution, freeing their mind from the chains of conformity and embracing the freedom of questioning. This introspective approach is not superficial; it has an existential depth that transcends the conventional limits of thought. Thus, the awakening of modern consciousness is an invitation to delve into the murky waters of existence and navigate beyond known horizons to discover new shores of reflection and perception. This inner awakening, deeply rooted in modernity, is proving to be the catalyst for an unprecedented psychological transformation, offering the soul the opportunity to embark on unexpected journeys towards regeneration.

The Inner Transformation: Thinking Beyond Dogma

For centuries, humanity has been shaped by a set of dogmas and beliefs that have dictated its perception of the world. However, amidst this tumultuous whirlwind of preconceived ideas, an inner transformation is taking shape, urging each individual to question established certainties. Thinking beyond dogma means embracing the unknown and exploring the dark corners of consciousness to uncover elusive truths.

This transformation urges us to free ourselves from the chains of intellectual conformity and let go of the preconceived judgements that burden our minds. The true act of rebellion lies in our ability to look beyond the limits imposed by tradition and embrace doubt as a faithful companion on our journey. It is through questioning that flashes of lucidity and bursts of liberated thought are revealed.

At the heart of this transformation, individuals confront their contradictions and the vastness of their inner universe. Deconstructing the foundations on which our certainties rest requires unparalleled courage and boldness of spirit to transcend the barriers of the known. It is in this space of uncertainty and assumed fragility that the seeds of authentic thought germinate, unhindered by inherited conceptual constraints.

Thinking beyond dogma also means opening ourselves to the wonders of human diversity and delving into the vast oceans of individual perspectives. We are not defined by narrow doctrines; we are the architects of our own worldview. Thus, inner transformation invites us to perpetual wonder and tireless discovery of the many facets of truth.

However, this paradigm shift is not without its challenges and disappointments. It involves a desperate quest: a journey through limbo where emotions, doubts, and desires collide. However, it is precisely in this precarious balance that the invisible threads of a new understanding and profound revelation are woven together. Inner transformation then becomes the vital gesture of a soul hungry for freedom — a frenzied dance towards the impenetrable horizon of the mind.

Science and Subjectivity: The Mind Under the Microscope

The influence of science on human subjectivity is a never-ending debate. Under the microscopic gaze of researchers, the mind appears to yield to objective truths, shedding its unruly trappings. Yet subjectivity persists, like an incorruptible spider's web, undeterred by repeated assaults by scientific objectivity.

The subtle mechanics of thought, observed through

scientific lenses, reveal treasures of infinite complexity. Each neuron and each synapse becomes a universe in its own right, a galaxy of mysterious interconnections where subjectivity takes root. Researchers venture into this neural labyrinth, deciphering electrical signals as poets interpret sonnets. Subjectivity is not dismissed out of hand, but revealed in all its elusive splendour.

However, the temptation to reduce subjective experience to mere brain chemistry remains strong. Eager to understand every behavioural phenomenon, neuroscientists search for traces of subjectivity in the most unexpected corners of the brain. However, despite their efforts, the spectre of subjectivity continues to haunt their models, refusing to be captured in the mesh of proteins and neurotransmitters.

The confrontation between science and subjectivity cannot be resolved in a sterile duel. On the contrary, it encourages us to explore the frontiers of objective knowledge and push back the boundaries of rational understanding. When subjected to microscopic scrutiny, subjectivity becomes an invitation to embrace the unknown, contemplate the imperceptible, and transcend the chains of fixed knowledge.

Fascinating though scientific truth may be, it cannot claim to be the sole expression of reality. By examining the mind under a microscope, science outlines not only the contours of grey matter, but also the translucent veils of subjectivity. Thus, rather than eclipsing our profoundly human sense of wonder, science offers a mirror in which the elusive brilliance of the mind is reflected.

The Simplicity of the Artistic Gesture: Truth or surrender?

In the depths of the human soul, the artistic gesture is an unparalleled manifestation of the perpetual quest for truth. It emanates from the very essence of existence, transcending the barriers imposed by objectivist thinking. The painter, hungry for meaning and authenticity, reveals themselves through the purity of their brushstrokes, exposing a fracture in the false fabric of the material world. Each stroke and nuance, whether striking or subtle, bears witness to a silent rebellion against the superficiality of an apparently unquestionable reality. Absorbed in their creation, artists free themselves from the chains of illusory certainties, reaching a state of grace where the beauty of their work is all that matters.

This raises the fundamental question: is this a surrender to the whims of emotion or privileged access to a deeper truth that transcends the rigid limits of rational knowledge? The apparent simplicity of the artistic gesture conceals an underlying complexity that can only be understood by those who dare to explore the depths of the human soul.

When the scientific and analytical gaze enters the intimacy of artistic creation, it encounters a reality that cannot be grasped by reason alone. The evocative power of certain works transcends usual appreciation criteria, defying critical thinking to touch viewers at the core of their being and shake the foundations of reflective, orderly perception. While science dissects the infinitely small, art deploys metaphorical and symbolic power to respond to the imperative to question our inner being and reveal the hidden mysteries of our human condition.

Therefore, the simplicity of the artistic gesture is neither absolute truth nor a defeat of rationality. Rather, it is an invitation to the human soul to explore the infinite possibilities of an elusive existence. By embracing this simplicity, the artist guides us towards a renewed consciousness where truth emerges in subtle nuances and harmonious contradictions.

Echoes of Free Thought: An Inevitable Metamorphosis

Art, an intimate reflection of the human spirit, is a breath of fresh air amid the monotony of everyday life. Like an echo resonating through the twists and turns of our thoughts, art frees us from our constraints and offers a fresh perspective on the complexities of our existence. The freedom of artistic creation acts as a catalyst, provoking a true inner metamorphosis in those who allow themselves to be carried away by its elusive currents.

Free thought blossoms like a rare flower within this social fabric, which is woven from influences and prohibitions. Breaking free from ideological constraints, it embraces its own essence and ventures towards unimagined horizons. The echoes of this rebellious thinking reverberate through every artistic masterpiece, sowing the seeds of transformation in the heart of the human soul.

Through art, individuals discover the infinite nuances of their own subjectivity. By rejecting pre-established norms, they confront their contradictions and reveal the richness of their inner duality. This confrontation stimulates reflection and invites deep introspection, marking the beginning of a true personal revolution.

The plurality of artistic voices resonates like a tumultuous symphony, prompting the mind to question its certainties and view reality from a fresh perspective. Art is no longer just an escape; it becomes a tool for self-reconstruction. It enables individuals to transcend the limits of their existence and access an unexpected fulfilment, transforming every shadow into a dazzling ray of light.

Through brushstrokes, emotionally charged verses and haunting melodies, the mind resolutely embarks on its own metamorphosis. Thus, artistic escape becomes the preferred means of exploring the depths of the soul, drawing individuals into a whirlwind of discoveries and existential questions. Under art's benevolent touch, the fragmentation of the ego transforms into captivating coherence, uniting every part of the self in a harmonious dance.

Free thought, nourished by the echoes of art, thus brings about a profound and inevitable metamorphosis within every soul hungry for freedom. Through this powerful fusion, individuals discover that they are the creators of their own reality, and that their life can be a magnificent work of their own mind, like the fullness of a finished painting.

Art as Escape: Fragmentation or fulfilment?

Art offers unconditional freedom to the mind in the throes of reality, providing an escape from the torments of existence. Like a divine craftsman, the artist weaves the threads of perception and emotion to create a parallel universe in which each person can find refuge. However, this artistic escape is not without psychological dilemmas. In its quest for fulfilment, the mind may find itself confronted with profound inner fragmentation, torn between earthly reality and the world of dreams.

The work of art then appears as a mysterious mirage, offering an alluring yet ambiguous refuge. Shimmering colours, enigmatic shapes and melodious words may invite transcendence, but they can also hide a cruel superficiality. Eager to escape, the viewer risks getting lost in this illusory labyrinth, never certain of finding the inner unity they seek. Thus, art oscillates between exalting the imagination and posing the danger of alienation, prompting human beings to navigate the infinite waters of the creative unconscious cautiously.

However, despite this inherent duality, art can also have a therapeutic dimension, providing an opportunity to gather the scattered fragments of the soul. Artistic creation then becomes an act of reconciliation and a means of transcending internal conflicts in order to achieve a new sense of fulfilment. From this perspective, art is not just an escape, but a path towards harmonious self-integration — an invitation to embrace all facets of our humanity, even the darkest ones.

Thus, art offers both an escape and a mirror, revealing the multiple facets of the human soul. Rather than being a mere distraction, art is the vehicle that enables us to navigate the complexities of our inner selves, striking a balance between fragility and strength, despair and hope. It is in this constant oscillation that the dazzling beauty of art lies, inviting us to reconcile opposites and contemplate fragmentation without losing sight of the fundamental unity of all existence.

Creative Duality: Essence and Existence in Human Action

The duality of humanity is a fundamental question that has spanned the ages, emphasising the conflict between essence and existence. It is the oh-so-famous gap between what we are and what we want to be. This duality creates a feeling of incompleteness and a perpetual quest for meaning and fulfilment. Yet it is precisely in this constant tension that humanity's creative force lies. It is through experiencing this duality that individuals find the impetus to transform their dreams into concrete actions. Creation, in whatever form, emerges from this inner dialogue — this tension between immutable essence and chaotic existence. The most powerful works and the most visionary ideas are born in this back-and-forth between ideal desire and tangible reality. Thus, the artist confronts their own duality, expressing their inner struggle through their creations and seeking to transcend the limits of their being in order to touch upon human universality. However, this duality is not confined to the artist alone. Everyone is the protagonist of their own masterpiece, weaving the fabric of their life from the threads of essence and existence. This constant inner dialogue fuels actions, shapes choices, and sculpts destinies. However, it is important not to fall into the trap of absolute dichotomy. It is in the harmony of these two opposing forces that true creativity flourishes. Confrontation with the brutality of existence should not result in the denial of one's deepest essence, but rather its elevation. Every act and every work of art is a synthesis of this duality — a complex reflection of the human condition. By recognising and embracing this duality, humanity can transcend its condition and reach new hights. Creative duality is the driving force behind metamorphosis and the root of all greatness. It is in the depths of the human soul that the epic of our existence unfolds like a symphony, dancing between essence and existence, shadow and light.

Chants Cérébraux: Words and Forms as Revelators

I invite you to delve deeper into the complex interaction between words and forms, guided by the flickering light of creative thought. Words are fragile yet indispensable links in human communication and carry immeasurable power within them. Their careful arrangement shapes ideas, shakes certainties and illuminates the dark corners of the soul. When they mingle with forms, whether on a canvas, in the structure of a sculpture or in the architecture of thoughts, they become silent messengers of our deepest essence. The forms themselves, through their silent language, reveal flashes of thought, capturing the echoes of emotion and inner turmoil. They embrace words, enveloping them in mystery and giving them unexpected depth. Thus, cerebral songs are born: graceful symphonies of thoughts and images intertwining to create new realities. Sometimes, words give way to artistic expression and forms are stripped bare to welcome the stories they tell. In this unique choreography, each element enriches the other, weaving a subtle network in which intellect dialogues with aesthetics and reason nestles in the sensual curves of creation. These cerebral songs carry us away in a whirlwind of fleeting emotions and eternal ideas, revealing the unexpected power of the alliance between words and forms. Defying time, they distil the quintessence of humanity through the centuries, offering the world an intellectual and spiritual caress that transcends the limits of reality. When words and forms merge in perfect harmony, the human spirit reaches its zenith, revealing the beauty of its duality and the infinite richness of its creative fervour.

Resonating with Oneself: The Arcane of Invisible Revolutions

The notions of resonance and revolution intertwine in a mysterious dance at the heart of the human soul. The modern collides with the eternal as it seeks to trace the winding path to the essence of the self. It is in this relentless quest that true upheavals take shape, hidden from the outside world. What is this inner alchemy that orchestrates our silent metamorphoses? The arcane of invisible revolutions lies in the depths of our being, where the echoes of our thoughts send shivers down our spine. The synergy of mind and soul gives rise to an unfathomable inner symphony, where each note resonates in harmony with the cosmic fabric. It is in this profound connection that we break free from the limitations of existence to embrace its universality. Beyond words and forms lies a universe of unspeakable meanings that subtly resonate with our essence. These invisible revolutions are born from the osmosis between rational will and creative intuition, where the self finds its inner voice and redraws the boundaries of consciousness. Resonating with oneself opens one up to the whispers of the unconscious, allowing one to explore the confines of one's subjectivity and discover the quintessence of one's being. Thus, introspective reflection becomes the crucible in which ideas and emotions converge to generate a new force that shakes the very foundations of the soul. Invisible revolutions weave invisible threads into an inner fabric, intertwining experiences and perceptions to create a complex web in which the evolving portrait of humanity emerges. Following these winding paths, the mind frees itself from the shackles of the ordinary to embrace the infinite richness of its own nature. In this mesmerising dance, transformation becomes the elusive link between science and art, reason and passion. This opens the door to a sublime synthesis, where the self ascends to new and unexpected horizons.

Towards a Sublime Synthesis: Charting the New Path of Being

There are moments in human life when a subtle transition takes place in silence, weaving the threads of existence to forge a new being. This inner genesis, often misunderstood and ignored, is nevertheless the very epicentre of our destiny. Thus begins the journey towards sublime synthesis: a quest to redefine the limits of humanity and transcend the barriers of conventional thinking.

At the heart of this journey lies a dance between art and science — a subtle orchestration of ideas and discoveries. These fields intertwine, influencing and transforming each other like inevitable partners in a cosmic ballet. In this existential waltz, human beings find themselves at a crossroads, confronted with their shadows and their light, their questions and their certainties.

Sublime synthesis is not merely the juxtaposition of disparate elements, but an intimate alchemy revealing the soul's essence. It evokes the harmonious fusion of opposites, blending the analytical rigour of science with the creative freedom of art and embracing intellect and emotion in a vibrant surge of unity.

However, this journey is not without its challenges, as it necessitates profound introspection and the re-evaluation of our deeply held beliefs and thought patterns. This transformation occurs in the dark recesses of the unconscious, where the essence of metamorphosis lies. It is a challenging path, filled with uncertainties and surprises, but it is also a transcendent adventure because it takes us beyond what we thought we knew.

By charting this new path, humanity ventures towards an infinite horizon and explores the uncharted territories of its own reality. Breaking free from the chains of conformity and banality, he creates his own destiny, liberated from the shackles that imprisoned his mind. Thus, the sublime synthesis spreads its majestic wings, inviting humanity to reach the pinnacle of its potential.

This new path of being is neither linear nor limited; it is strewn with questions and discoveries, woven from dreams and challenges. It stretches out before us like a blank canvas, ready to welcome the brushstrokes of our creative essence. This path is one of rebirth, self-revelation and the ultimate fulfilment of our humanity. The sublime synthesis thus traces a luminous path through the twists and turns of existence, offering humanity the promise of a radiant metamorphosis woven from the infinite threads of knowledge and art.

10

Towards Harmonious Coexistence

Science, art and religion in perpetual dialogue

The inseparable brothers of knowledge: Science and Art

The indissoluble union between science and art is a fascinating enigma that has sparked the interest of curious minds throughout the centuries. Despite being often perceived as opposites, these two disciplines actually maintain a deeply symbiotic relationship that transcends the dogmatic boundaries of their own fields. The meticulous exploration of natural phenomena and the mysteries of the universe by scientists is akin to the creative process of artists seeking to represent and interpret the world around them. Indeed, art and science converge in their shared pursuit of truth, beauty, and human understanding. While science dissects phenomena to reveal the laws that govern the universe, art draws on aesthetics to express the emotions and inner experiences that define the essence of humanity. The fruitful synergy between science's analytical rigour and art's creative freedom not only enriches our view of the world, but also nourishes our inner being. The bridges woven between these two disciplines offer unique perspectives that illuminate the human condition. With tireless intrepidity, scientists explore the arcane workings of the universe, unveiling its most intimate mysteries, while artists probe the depths of the human soul, revealing its emotional complexity with an almost divine force. It is in this harmonious interplay of scientific rigour and artistic intuition that humanity finds a path to transcendent knowledge and immortal beauty.

Spirituality Revealed Through the Prism of Science

Rather than being opposed, art and science resonate with each other to capture the most subtle vibrations of the soul. By examining the infinite universe through the enlightened perspective of the scientist, a symphony of mysterious forces is revealed that transcends the limits of matter. It is in this desperate quest for truth, beyond formulas and equations, that spirituality enters the temple of rational knowledge. Like modern alchemists, scientists explore the mysteries of existence with fervour, revealing the divine essence that permeates every atom of this immeasurable universe.

Thus, contemplation of nature opens the doors to meditation, and scientific research becomes an opportunity for communion with the most elusive mysteries of existence. After all, who better than the scientist, immersed in the infinite ocean of knowledge, can perceive the presence of a transcendent creative force? Every discovery and every advance in understanding the laws that govern the universe reveals a fragment of the ultimate wisdom that weaves the invisible threads of all reality. The prism of the scientist thus becomes a mirror reflecting the eternal light of spirituality, offering the world a renewed vision of its deepest essence.

Similarly, the artist, with their delicate sensitivity,

guides us towards the ineffable. Through their creations, they invite us to contemplate infinity and feel the vital force pulsing at the heart of all things. Their works transcend the barriers of space and time, touching our inner being and awakening a deep resonance with the entire universe within us. Thus, a timeless dialogue unfolds between reason and intuition, between Cartesian logic and the poetry of inspiration, from the eye of the scientist to the hand of the artist.

This encounter between science, art, and spirituality is much more than mere coexistence. It is a celestial ballet in which each dancer — whether a scientist, an artist, or a person of faith — brings their own unique contribution to the cosmic symphony. It is in this universal harmony that the promise of revelation lies: a truth that transcends human categories and rises above contingency. Thus, the scholar's prism becomes the subtle echo of the universal soul, like a divine song that unites humanity in its desperate quest for meaning and beauty.

When art transcends the limits of pure reason

Art, a subtle manifestation of the human spirit, rises above the narrow boundaries of pure reason. It transcends the limits of our empirical understanding to reveal the ineffable truths of existence. In every painting, sculpture and musical composition lies a captivating enigma that defies rationalism. The artist becomes a daring guide, exploring the depths of the soul and attempting to unveil its most intimate mysteries.

When a work of art engages our senses, it transports us to a realm where pure logic dissolves in the face of enigmatic beauty. Forms, colours and sounds are all transformed into a universal language that evokes indescribable emotions and elusive thoughts. It is in this transcendence that the true power of art is revealed, opening the doors of perception and feeding the minds of those who hunger for deeper meaning.

Poetry, for instance, defies the limitations of straightforward rational analysis. It weaves words into images, feelings, and life experiences. It gives the human soul the chance to venture beyond the calm waters of logic and plunge into the turbulent depths of raw emotion. Similarly, dance transcends the limits of reason by expressing stories without words, emotions without speech and truths without mathematical formulas.

Thus, by transcending the contingencies of pure thought, art reveals essential truths that elude all attempts at rationalisation. Art encourages us to explore the uncharted territories of the imagination and contemplate the infinite within every artistic creation. In this way, art joins science and religion in their shared quest to understand the world. Art offers a new vision and an extra dimension to our human experience by embracing the irrational in order to better grasp the complexity of our condition.

Faith and Logic: A Dialogue at the Heart of Man

The question of faith and logic is a major issue in the quest for meaning that has driven humanity since its origins. Faith, based on spiritual or religious beliefs, sometimes seems to clash head-on with the rigour of rational logic. This apparent incompatibility has fuelled much reflection and debate throughout history. However, it seems that not only is the coexistence of these two pillars of human thought possible, it is also necessary for a holistic understanding of existence. Although faith is often considered irrational, it also manifests itself through intuition, lived experience and an emotional connection with the sacred. It is a different source of knowledge, belonging to a logic that is more spiritual than rational. Conversely, logic provides valuable tools for understanding how the world works, analysing our perceptions, and interacting with our environment. Recognising the roles of faith and logic in our lives enables us to adopt a more comprehensive view of reality, where intuition and reason coexist and spirituality enriches our perception of the material world. Recognising the ongoing dialogue between faith and logic enables us to transcend dualities and appreciate the depth of the human experience. Rather than opposing each other, these two dimensions complement each other, leading to inner harmony and offering an enlightened perspective on existence.

Harmony or discord? The Game of Individual Perspectives

At the heart of the human condition lies the eternal debate between harmony and discord — an omnipresent dialectic that transcends time and space. Individual perspectives often diverge, creating a complex interplay where resolution eludes understanding. Reality is never absolute, but modulates according to personal beliefs and experiences. Every person carries within them a reflection of their own subjectivity, which colours their view of the world with a unique hue. This infinite diversity gives rise to a kaleidoscope of opinions, each fragment shining with its own light. However, beyond this apparent fragmentation, a mysterious beauty emerges from the whole, like a work of art composed of multiple disparate yet harmoniously arranged elements. In this ocean of infinite perceptions lies the potential for deep understanding, where the peaceful coexistence of ideas provides fertile ground for intellectual and spiritual growth. It is through the ongoing dialogue between these perspectives that the fabric of human richness is woven, revealing the complexity and nobility of the human spirit. Thus, apparent discord conceals an exhilarating polyphonic symphony, offering humanity the chance to rise above temporary disharmony and reach the heights of mutual understanding and collective enrichment.

Artistic creation is a bridge between the sacred and the profane

Art, that marvellous alchemy of the human soul, reveals itself to be the crucible in which the sacred and the profane converge. A painted canvas, a symphony performed, a poem recited — these are all manifestations that transcend the earthly condition and strive towards the infinite. Is the artist not a true demiurge, shaping and moulding raw material to reveal its divine brilliance? Each brushstroke, musical note and carefully chosen word resonates as an offering to humanity, inviting us to contemplate the immeasurable beauty of the universe.

By its very nature, art opens the doors of perception, inviting us to explore the deepest territories of the soul. Thus, it builds a bridge between the immanence of the tangible world and the transcendence of the celestial spheres. In this delicate equilibrium, the artwork becomes the very embodiment of the sacred within the mundane, providing a cherished escape to a timeless realm where the boundaries between the material and spiritual become indistinct.

From cave paintings to museum masterpieces, art has always been a privileged means of expressing the inexpressible and giving form to the unfathomable. Through artistic creation, humanity explores and celebrates its own condition, weaving mysterious links between heaven and earth. From this perspective, art unfolds as a perpetual quest for meaning and transcendence, magnifying and sublimating reality.

Ultimately, by building bridges between the sacred and the profane, artists offer humanity the precious gift of reconnecting its deepest aspirations with sublime cosmic energy. Like a tightrope walker balancing on the thread of existence, the artist guides us towards reconciling our earthly nature with our aspiration for the infinite, creating a permanent dialogue between humanity and the divine.

Echoes of Light: The Word and the Verb in the Universal Tribune

The luminous echo of the word and the verb resonates through the ages, transcending the boundaries of time and space. In this universal arena, the evocative power of words weaves a golden thread between the sacred and the profane, illuminating the human soul with its eternal radiance. Words unfold like celestial constellations, illuminating the mind with their divine breath and bearing truth. Each syllable becomes a beacon in the darkness of troubled minds, opening the way to a deeper understanding of the universe and of oneself. In its purest form, speech is an offering to reflection, an intoxicating dance between the lines of silence and the meanders of meaning. Meanwhile, words sculpt reality with the precision of a goldsmith, bringing to life the worlds hidden in the depths of the imagination.

In this universal forum, verbal exchange becomes a cosmic ballet where thoughts converge and intertwine like moving stars. Every word, every idea and every conviction finds an echo in another's heart, creating a harmonious symphony within the mind. Words become rays of light, piercing the darkness of solitude and uniting beings in spiritual communion that transcends material boundaries. This universal forum offers a refuge for thinkers in search of intellectual camaraderie, where diverse opinions come together in an exhilarating dance to celebrate the infinite richness of human thought.

Thus, speech and words rise like resplendent phoenixes, spreading their majestic wings above earthly contingencies. Their power transcends cultures, beliefs, and eras, building immaterial bridges between civilisations and opening up new horizons for humanity. In this universal forum, words rise like a hymn to freedom of spirit while language acts as a mirror to the mind, reflecting the beauty and complexity of the human experience. The echoes of light thus resonate, acting as ambassadors for the universal soul and inviting everyone to listen, meditate and share in this vast mosaic of thoughts, emotions and visions.

A Cosmic Dance: Synchronicity of Souls and Ideas

In the vast expanse of the cosmos, human souls embark on an eternal dance, seeking harmony and unity beyond physical boundaries. Ideas, like distant stars, converge and diverge in an endless ballet, revealing the intricacy of the human experience. As minds intertwine, a subtle synchronicity emerges, transcending the limits of individual understanding and awakening a collective consciousness. Though silent to the material world, this cosmic dance resonates in the depths of every being, inspiring an endless quest for meaning and truth. Like a mirror reflecting the brilliance of constellations, each soul finds its place in this universal symphony and makes a unique contribution to the whole. Ideas evolve like moving planets, forming and challenging connections in a fragile yet vital balance. This synchrony of souls and ideas transcends the boundaries of duality, weaving an indissoluble fabric of mutual connections. In this dance, human thought becomes the beacon that guides our aspirations towards unexplored horizons, embracing the mystery and beauty of existence. Every encounter, every exchange of ideas and every creative gesture contributes to building this celestial orchestra, where harmony does not reside in uniformity but in the magnificent diversity of unique voices. Thus, the human soul merges into this cosmic dance, discovering the richness of plurality and the resonance of differences. This synchrony of souls and ideas invites us to listen to the universe's timeless music, share our thoughts like luminous constellations and celebrate the fundamental unity that binds all of humanity in an infinite dance of discovery and creation.

Towards the Awakening of a New Human Sensibility

The last century has witnessed unprecedented upheavals in collective consciousness. Scientific advances, artistic revolutions, and religious movements have shaped a constantly evolving cultural landscape. However, beyond the disciplinary boundaries that separate them, a common quest is emerging: the quest for the essential and for universality in our human experiences.

Inexorably, the evolution of ideas brings together disciplines that seem diametrically opposed. Science, once the guardian of absolute truth, is merging with the metaphysical questioning of art, while religion is undergoing a transformation through the lens of objective knowledge. Rather than diluting the essence of each discipline, this symbiotic process enriches our understanding of the world and opens up unexpected horizons.

At the same time, human sensibility, tormented by the upheavals of progress, is evolving too. In an era of triumphant individualism that threatens to divide humanity, a new sensibility is slowly emerging — the product of the symbiosis between science, art, and spirituality. It embraces the diversity of human experience, transcends fixed dogmas, and unites hearts across divides.

In this era of critical thinking, the emergence of this new sensibility could be the gateway to greater mutual understanding. It encourages us all to listen to one another kindly, to contemplate nature's wonders and to allow ourselves to be moved by the indescribable beauty of artistic expression. In the face of contemporary challenges, this sensibility offers hope for a more united humanity that is capable of addressing the great problems of our time with empathy and intelligence.

As science, art and religion continue their infinite dialogue, a new human sensibility is being patiently built, weaving the threads of a broader vision of existence. In this rediscovered fraternity, humanity may find the key to opening the doors to a more harmonious world, shaped by compassion, intelligence, and creativity.

The foundations of a world connected by the thread of reflection

Over the centuries, humanity has oscillated between the extremes of knowledge, tirelessly seeking to unravel the mystery of life. Scientific discoveries have illuminated our understanding of the material world, revealing wonders and challenging our perception of reality. At the same time, art has been the witness to the deepest expression of the human soul, offering an intimate view of the human experience throughout the ages. And beyond these explorations, religion has continued to offer spiritual guidance, fuelling the flame of faith and providing comfort in times of doubt. Thus, this trinity of science, art and religion weaves the invisible threads that unite humanity. In this complex web woven by the foundations of human thought, a map of the mind emerges, revealing the subtle connections between the different facets of our existence. It is in this creative ferment that a new human sensibility is beginning to emerge, giving birth to a world where reflection becomes the cornerstone of society. Far beyond cultural and religious boundaries, this universal resonance transcends conventional paradigms to open up a space for unlimited dialogue, where interdisciplinarity becomes the keystone of human evolution.

Thus, we find ourselves at the dawn of an era where reflection, illuminated by the light of knowledge, the passion of art, and the wisdom of spirituality, is forging the foundations of a world that is not divided, but connected by an invisible thread of mutual understanding. By embracing this global vision, humanity can hope to achieve a coexistence where differences are no longer synonymous with conflict, but with shared wealth. This world, underpinned by reflection and introspection, offers a landscape where each individual finds their place, not despite their beliefs, but thanks to the diversity of perspectives they help to weave. Thus, it is in the foundations of this shared reflection that the path to true harmony opens up, shaped by the fusion of knowledge, emotions and convictions. And it is in this interconnected fabric of thought that the future of humanity is taking shape, forged by the thread of reflection.

Paris, le 14 July 2020.

www.ingramcontent.com/pod-product-compliance
Lightning Source LLC
LaVergne TN
LVHW090408210926
848480LV00002B/500